D1109961

Q
158.5
N38
1986.

PHYSICS THROUGH THE 1990s

Scientific Interfaces and Technological Applications

Panel on Scientific Interfaces and Technological Applications

Physics Survey Committee

Board on Physics and Astronomy

Commission on Physical Sciences, Mathematics, and Resources

National Research Council

NATIONAL ACADEMY PRESS
Washington, D.C. 1986

NATIONAL ACADEMY PRESS 2101 Constitution Avenue, NW Washington, DC 20418

NOTICE: The project that is the subject of this report was approved by the Governing Board of the National Research Council, whose members are drawn from the councils of the National Academy of Sciences, the National Academy of Engineering, and the Institute of Medicine. The members of the committee responsible for the report were chosen for their special competences and with regard for appropriate balance.
This report has been reviewed by a group other than the authors according to procedures approved by a Report Review Committee consisting of members of the National Academy of Sciences, the National Academy of Engineering, and the Institute of Medicine.

The National Research Council was established by the National Academy of Sciences in 1916 to associate the broad community of science and technology with the Academy's purposes of furthering knowledge and of advising the federal government. The Council operates in accordance with general policies determined by the Academy under the authority of its congressional charter of 1863, which establishes the Academy as a private, nonprofit, self- governing membership corporation. The Council has become the principal operating agency of both the National Academy of Sciences and the National Academy of Engineering in the conduct of their services to the government, the public, and the scientific and engineering communities. It is administered jointly by both Academies and the Institute of Medicine. The National Academy of Engineering and the Institute of Medicine were established in 1964 and 1970, respectively, under the charter of the National Academy of Sciences.

The Board on Physics and Astronomy is pleased to acknowledge generous support for the Physics Survey from the Department of Energy, the National Science Foundation, the Department of Defense, the National Aeronautics and Space Administration, the Department of Commerce, the American Physical Society, Coherent (Laser Products Division), General Electric Company, General Motors Foundation, and International Business Machines Corporation.

Library of Congress Cataloging-in-Publication Data

National Research Council (U.S.). Physics Survey Committee.
Scientific interfaces and technological applications.

(Physics through the 1990s)
Includes index.
1. Science. 2. Physics. 3. Technology.
I. Title. II. Series.
Q158.5.N38   1986   500   85-32039
ISBN 0-309-03580-5

Printed in the United States of America

First Printing, April 1986
Second Printing, September 1986

PANEL ON SCIENTIFIC INTERFACES AND TECHNOLOGICAL APPLICATIONS

WATT W. WEBB, Cornell University, *Co-Chairman*
PAUL A. FLEURY, AT&T Bell Laboratories, *Co-Chairman*
TED G. BERLINCOURT, Office of Naval Research
AARON N. BLOCH, EXXON Research and Engineering Company
ROBERT A. BUHRMAN, Cornell University
PETER EISENBERGER, EXXON Research and Engineering Company
MITCHELL FEIGENBAUM, Cornell University
KENNETH A. JACKSON, AT&T Bell Laboratories
ANGELO A. LAMOLA, Polaroid Corporation
JAMES S. LANGER, University of California, Santa Barbara
ROBERT L. PARK, University of Maryland
WILLIAM H. PRESS, Harvard University
DONALD TURCOTTE, Cornell University
ALEXANDER ZUCKER, Oak Ridge National Laboratory

Consultants

LEE A. BERRY, Oak Ridge National Laboratory
MARC H. BRODSKY, IBM Research Center
BRITTON CHANCE, The University of Pennsylvania
JACK DAVID COWAN, The University of Chicago
PETER G. DEBRUNNER, University of Illinois
FRANCIS J. DI SALVO, JR., AT&T Bell Laboratories
ELLIOT ELSON, Washington University, St. Louis
DAVID K. FERRY, Arizona State University
HANS FRAUENFELDER, University of Illinois
BYRON B. GOLDSTEIN, Los Alamos National Laboratory
WARREN D. GROBMAN, IBM Research Center
DANIEL L. HARTLEY, Sandia National Laboratories
JOHN J. HOPFIELD, California Institute of Technology/AT&T Bell
    Laboratories
ANDREW KALDOR, EXXON Research and Engineering Company
MARK H. KRYDER, Carnegie Mellon University
AARON LEWIS, Cornell University
L. K. MANSUR, Oak Ridge National Laboratory
WOLFGANG PANOFSKY, Stanford University
ALAN PERELSON, Los Alamos National Laboratory
PHILIP PINCUS, EXXON Research and Engineering Company
FRANCES E. SHARPLES, Oak Ridge National Laboratory

iii

CHARLES STEVENS, Yale University
JOHN C. TULLY, AT&T Bell Laboratories
EDWARD WITTEN, Princeton University
RICHARD F. WOOD, Oak Ridge National Laboratory

iv

PHYSICS SURVEY COMMITTEE

WILLIAM F. BRINKMAN, Sandia National Laboratories, *Chairman*
JOSEPH CERNY, University of California, Berkeley, and Lawrence
Berkeley Laboratory
RONALD C. DAVIDSON, Massachusetts Institute of Technology
JOHN M. DAWSON, University of California, Los Angeles
MILDRED S. DRESSELHAUS, Massachusetts Institute of Technology
VAL L. FITCH, Princeton University
PAUL A. FLEURY, AT&T Bell Laboratories
WILLIAM A. FOWLER, W. K. Kellogg Radiation Laboratory
THEODOR W. HÄNSCH, Stanford University
VINCENT JACCARINO, University of California, Santa Barbara
DANIEL KLEPPNER, Massachusetts Institute of Technology
ALEXEI A. MARADUDIN, University of California, Irvine
PETER D. MacD. PARKER, Yale University
MARTIN L. PERL, Stanford University
WATT W. WEBB, Cornell University
DAVID T. WILKINSON, Princeton University

DONALD C. SHAPERO, *Staff Director*
ROBERT L. RIEMER, *Staff Officer*
CHARLES K. REED, *Consultant*

v

BOARD ON PHYSICS AND ASTRONOMY

HANS FRAUENFELDER, University of Illinois, *Chairman*
FELIX H. BOEHM, California Institute of Technology
RICHARD G. BREWER, IBM San Jose Research Laboratory
DEAN E. EASTMAN, IBM T.J. Watson Research Center
JAMES E. GUNN, Princeton University
LEO P. KADANOFF, The University of Chicago
W. CARL LINEBERGER, University of Colorado
NORMAN F. RAMSEY, Harvard University
MORTON S. ROBERTS, National Radio Astronomy Observatory
MARSHALL N. ROSENBLUTH, University of Texas at Austin
WILLIAM P. SLICHTER, AT&T Bell Laboratories
SAM B. TREIMAN, Princeton University

DONALD C. SHAPERO, *Staff Director*
ROBERT L. RIEMER, *Staff Officer*
HELENE PATTERSON, *Staff Assistant*
SUSAN WYATT, *Staff Assistant*

COMMISSION ON PHYSICAL SCIENCES, MATHEMATICS, AND RESOURCES

HERBERT FRIEDMAN, National Research Council, *Chairman*
THOMAS D. BARROW, Standard Oil Company (Retired)
ELKAN R. BLOUT, Harvard Medical School
WILLIAM BROWDER, Princeton University
BERNARD F. BURKE, Massachusetts Institute of Technology
GEORGE F. CARRIER, Harvard University
CHARLES L. DRAKE, Dartmouth College
MILDRED S. DRESSELHAUS, Massachusetts Institute of Technology
JOSEPH L. FISHER, Office of the Governor, Commonwealth of
  Virginia
JAMES C. FLETCHER, University of Pittsburgh
WILLIAM A. FOWLER, California Institute of Technology
GERHART FRIEDLANDER, Brookhaven National Laboratory
EDWARD D. GOLDBERG, Scripps Institution of Oceanography
MARY L. GOOD, Signal Research Center
J. ROSS MACDONALD, University of North Carolina
THOMAS F. MALONE, Saint Joseph College
CHARLES J. MANKIN, Oklahoma Geological Survey
PERRY L. MCCARTY, Stanford University
WILLIAM D. PHILLIPS, Mallinckrodt, Inc.
ROBERT E. SIEVERS, University of Colorado
JOHN D. SPENGLER, Harvard School of Public Health
GEORGE W. WETHERILL, Carnegie Institution of Washington

RAPHAEL G. KASPER, *Executive Director*
LAWRENCE E. MCCRAY, *Associate Executive Director*

vii

Preface

Physics traditionally serves mankind through its fundamental discoveries, which enrich our understanding of nature and the cosmos. While the basic driving force for physics research is intellectual curiosity and the search for understanding, the nation's support for physics is also motivated by strategic national goals, by the pride of world scientific leadership, by societal impact through symbiosis with other natural sciences, and through the stimulus of advanced technology provided by applications of physics.

This Physics Survey volume looks outward from physics to report its profound impact on society and the economy through interactions at the interfaces with other natural sciences and through applications of physics to technology, medicine, and national defense. Six other volumes recount the status, progress, and promise of physics as scientific enterprise in the major physics subfields of particle physics; nuclear physics; plasma physics and fluids; condensed-matter physics; atomic, molecular, and optical physics; and cosmology, gravitation, and cosmic-ray physics. An overview volume addresses the overall progress, opportunities, and needs of the field; analyzes the status and trends of U.S. physics on the world scene; and discusses funding, manpower, and demographic issues.

The vital role of physics in propelling the entire scientific enterprise and in providing the basis for new technologies of immense economic importance is evident throughout this Survey, but it is the present volume that focuses on technological applicability.

New scientific disciplines are arising from the scientific interfaces between physics and biology, geology, and materials science. The traditional interfaces with chemistry and mathematics are acquiring new connections that promise revolutions in the way that research is conducted. These scientific interfaces are explored in 6 chapters that illustrate selected examples of the progress and trends that appear in appropriate symbiotic relationships. The emerging impression is a scientific whole that greatly exceeds the sum of its parts. Interdisciplinary science at the most fundamental research level is shown to be one of the most exciting promises for the coming decade.

The interfaces between physics and the various engineering disciplines are not treated explicitly. But the crucial continuum of physics activity from the most basic research to engineering applications is emphasized. Its reality must be recognized, supported, and strengthened for the national good. Indeed, more people trained in physics are engaged in applications or engineering than in pure research or academic pursuits.

Microelectronics, optical communications, and new instrumentation exemplify the major technological applications of contemporary physics research. A diversity of physics applications contributes to advances in medical diagnosis and treatment, to national defense systems, and to solutions of energy and environmental problems. Six chapters of this volume discuss these important areas of application of physics, largely in terms of their societal and economic impact.

Consideration of the interfaces and applications of physics leads naturally to recommendations aimed at strengthening the vital continuum joining science, technology, and our national prosperity. They are specified and explained at the conclusion of Chapter 1, Summary and Recommendations.

Contents

Executive Summary

Physics research combines a drive for understanding of the natural universe with a pervasive influence on the quality of human life through the technology that it generates. The exhilarating twentieth-century pace of discovery and application has accelerated in the past decade. New technologies, such as high-speed electronics, optical communication, advanced medical instrumentation, exotic defense systems, and energy and environmental systems, have nucleated and grown to maturity within only a few years after the physical discoveries on which they are based. At the same time, new ideas and methods born at the scientific interfaces are evolving scientific capability to address ever more complex problems. Physics has both contributed to and drawn from the sciences of chemistry, biology, and mathematics in entirely new ways. Computer science, geology, engineering, and materials science have become fully symbiotic with physics, to the benefit of humankind.

This volume illustrates these advances through representative examples and recommends actions to enhance the value of physics to society through both its scientific interfaces and its technological applications. Recognizing the importance and complexity of the science-technology-economy continuum, it seeks to strengthen these interconnections in key areas.

In addition to general recommendations put forth in the Overview volume of the Physics Survey, we propose the following:

- Funding agencies should devise procedures to evaluate and support interdisciplinary research collaborations involving participants from deep within the associated disciplines. Special attention should be given to start-up research grants for young faculty to begin interdisciplinary programs.

- The universities should promulgate interdepartmental, interdisciplinary research programs and centers to transcend barriers among the traditional disciplines, to provide transdisciplinary education, and to attract and utilize interdisciplinary research funding.

- Universities and funding agencies should organize to accommodate and enhance the engineering-physics interface in both education and research.

- The federal government should encourage in-house industrial fundamental research and cooperative industrial research with universities, national laboratories, and other industry through appropriate tax and antitrust policies.

- The support of long-range fundamental research should be shared by the mission-oriented agencies, particularly the U.S. Department of Defense, and be restored to at least pre-Mansfield Amendment (1970) levels.

1

Summary and Recommendations

PERSPECTIVES FOR SOCIETY: APPLICATIONS, IMPLICATIONS, AND INTERFACES OF PHYSICS

Research in physics increases our fundamental understanding of nature, generating knowledge with far-reaching consequences for humankind. The new technology that it has spawned is so ingrained in our civilization that its scientific origins are often overlooked. This volume of *Physics Through the 1990s* reports on the profound societal impact of physics through its applications in technology and its interfaces with other sciences.

As a fundamental science, physics presents some of the deepest challenges ever to engage the human mind. The excitement of discovery in physics has accelerated its pace throughout this century. The other volumes of this survey detail these discoveries in the subfields of elementary particles; plasmas and fluids; nuclear as well as atomic, molecular, and optical physics; condensed matter; and cosmology and gravitation.

Unlike the sometimes altruistic patronage of the arts by medieval authorities, the substantial level of contemporary federal support of physics research is motivated by the anticipation of practical benefits. Yet despite media accounts of spectacular discoveries in such areas as elementary-particle physics, interplanetary science, nuclear fusion, and exotic superconductors, scientific research often seems to remain

3

disconnected in the public mind from practical concerns of the U.S. economy. This volume demonstrates the value of physics to society and conveys a sense of the importance to the economic health of the United States of its world leadership in both fundamental research and technological innovation.

From the breadth of physical activities that compose the applications of or scientific interfaces with physics, this volume reports on several fields of current excitement and future potential, as well as several fields of immediate importance to society. The selected interfaces represent those in which interdisciplinary interaction is particularly vigorous at fundamental levels: biophysics, materials science, the chemistry-physics interface, geophysics, and mathematical and computational physics.

From among the innumerable applications of physics, we have selected several areas that combine pivotal dependence on recent research with an identifiable large-scale industrial technology. We highlight applications of physics in electronics, in optical information technologies, in new instrumentation for both science and society, in the fields of energy and environment, in national security, and in medicine.

In many of these applications the lines between physics and engineering are frequently and necessarily crossed. Although engineering is not treated explicitly within this volume, its crucial role permeates essentially every aspect of the technologies represented here. Continued technological progress depends on intense interaction between physics and engineering just as continued scientific progress depends on interactions between physics and other sciences. Indeed, the continuum of viewpoints shared by science, technology, and engineering is a major underlying theme of this volume.

In each of the chapters the major advances of the past decade are surveyed within the framework of physics research, focusing on representative examples rather than exhaustive compilations. Specific recommendations of a particular technology or scientific interface appear in the appropriate chapters. Recommendations that transcend individual specialties and suggest programs by which society and science may continue to benefit from progress in physics are drawn together in the final section of this chapter.

SCIENTIFIC SYNERGY: THE SCIENTIFIC INTERFACES OF PHYSICS

Because the principles of physics serve as a foundation for other sciences, new discoveries in physics frequently stimulate research in associated sciences. In turn, problems of other sciences can present physicists with profound challenges.

Each of the 6 interfaces presented here illustrates different implications of physics, presents its own opportunities, and leads to particular conclusions and recommendations.

Although the interfaces of physics with the various engineering disciplines have not been addressed here explicitly, we note emphatically that there exists in physics a continuum of technical activities without any sharp distinctions from the most fundamental scientific research to the most immediate technological applications. Thus, in those areas of technology where new physical science is moving rapidly into application, there is no point in attempting to distinguish between physics and engineering.

Biological Physics

The elegant complexity of life often diverts our attention from its ultimate reliance on the principles of physics. The tools of physics have always been involved in biological research, and the fundamentals of biological phenomena have increasingly challenged physicists. Early optical microscope observations of bacterial motion (Brownian motion) led Albert Einstein to formulate the basic statistical mechanics of fluid diffusion, and it was modern electron microscopy that identified the structure of molecular photodetectors packed in a two-dimensional lattice of cell membrane.

Biophysics is, however, more than the application of physical methods to biological systems. Fundamental biological challenges stimulate new applications of concepts of theoretical physics. Today, for example, the *biopolymers*—the proteins and genetic materials DNA and RNA—have a well-established empirical chemistry thanks to continuously improving physical experiments. Fundamental understanding of protein structure and function now appears to be a realistic research objective. Quantum-mechanical calculations of structure and dynamics in terms of molecular fluctuations are beginning to reach a satisfactory level of approximation with the help of models, techniques, and computational capabilities derived from physics.

In both *cell physiology* and *neurobiology*, physical methods have long been crucial to understanding molecular transport, membrane structure, and signal processes in the brain, nerves, and muscles. New physical measurements of ion currents through single transmembrane molecular channels have provided the first direct access to the central molecular mechanisms that govern signaling processes in the brain and nerves.

Gene manipulation biotechnology now makes possible the systematic modification and production of potentially any protein. Thus, for the first time, rare proteins can be produced in quantities sufficient for structural studies by physical and chemical methods, and the expression of natural or modified proteins in living cells can be controlled to identify structure-function relations. The combination of biotechnology and biophysics offers a splendid promise for future productive research.

The multicellular *organization of the brain* and its principles of information storage and retrieval pose fundamental questions about physical networks that are currently attracting the attention of mathematical physicists. The theoretical physics of partially ordered systems has recently been applied with dramatic success to model aspects of the brain function. This could lead to significant interdisciplinary synergism involving the fields of neuroscience, statistical physics, and computer science, with implications extending to robotics and artificial intelligence. Perhaps ultimately the understanding needed to treat organic failures of the brain will emerge.

One might say that physics has advanced to a level on which it can now begin to address the imposing complexity of fundamental biological science at both molecular and many-body organizational levels. Yet institutional organization of both education and funding for biological physics often seems to impede progress at this fruitful interface. Few university physics departments accommodate biophysics. Biophysics is incorporated effectively in many biological science research programs, but the cross-fertilization between modern physics and biology does not often seem to be adequately supported. We can point to an opportunity here for effective interdisciplinary interaction through the funding of interdisciplinary centers that parallels that initiated so effectively in materials science 20 years ago.

Physics and Materials Science

In the past three decades, materials science has developed as an independent discipline from a fusion of metallurgy, chemistry, and ceramics engineering with aspects of condensed-matter physics. Today, at their common boundary, the fields of materials science and condensed-matter physics are distinguishable primarily by the viewpoints of their complementary educational disciplines.

The interface between physics and materials science is continuously innovative. The spectrum from basic problems to applications in technology is truly continuous. The challenge of problems in materials science attracts the theorists, while experimentalists respond enthusiastically to new compositions and conditions of matter provided by new technology. Research in new materials, processes, and analytical methods involves many areas of physics. Ion-beam etching, sputtering, and molecular-beam interactions with surfaces involve plasma physics and atomic physics. Analytical methods in materials science include those of atomic physics, such as Rutherford scattering and particle beams. As statistical physics has succeeded in the analysis of partially ordered many-particle systems, the understanding of glasses and amorphous materials has advanced. The amorphous state is a topic of great current interest, attracting investigators with both fundamental and applied motivations. The dynamical theory of nonlinear systems is an extremely lively topic in mathematical physics that has immediate applications to important materials-science problems, including crystal-growth instabilities and dynamic structure modulation.

Interdisciplinary research centers in universities have proven remarkably effective in initiating and sustaining progress at the materials-science/physics interface. The large industrial research laboratories also provide a primary force for progress at this interface since their most valuable successes are typically based on interdisciplinary research. It is important to note that the condensed-matter-physics/materials-science interface calls mainly for small-group research, in which individual investigators are the source of ideas and progress. But lack of funds for equipment and instrumentation to provide state-of-the-art technology for materials fabrication, processing, and analysis often retards progress in academic laboratories. Sophisticated facilities in the million-dollar range (such as molecular-beam or electron-microscopy apparatus) are required at numerous locations to carry out this productive small-group research. While research centers can efficiently provide advanced-technology facilities, it is also appropriate to support smaller user groups that can share facilities.

The Physics-Chemistry Interface

The interface between chemistry and physics is enduring but constantly changing. Physical chemists and chemical physicists are often hardly distinguishable. Recently, the interdisciplinary associations have begun to reach more deeply into both fields, as when, for example, synthesizing chemists collaborate with atomic and optical physicists.

The chapter in this volume on the physics-chemistry interface treats several topics of broad current interest and technological applicability and omits, without prejudice, many other fundamental topics of chemical physics that have been discussed in other volumes of this Survey or in the recent chemistry survey (*Opportunities in Chemistry*, National Academy Press, Washington, D.C., 1985). Emphasized here are laser chemistry, surface science, neutron and x-ray techniques, polymers and complex fluids, and electrically conducting organic materials.

The development of tunable lasers has made possible spectroscopies of spectacular sensitivity as well as exciting new strategies for observation of unstable and excited molecular states. The technologies of molecular beams, transient surface states, optically driven chemical processes, and laser excitation for isotope separation have all developed from applications of lasers to chemical physics.

Surface science is involved in some of the most important and puzzling of chemical processes. Heterogeneous catalysis involves profound interactions between surfaces and the chemical reactions taking place upon them. These quasi-two-dimensional interfaces pose fundamental challenges to both theory and experiment. The most powerful tools are used to study the nature of surfaces, the states of absorbed molecules, and the transient products of surface reactions.

While the fluids formed by simple molecules have come to be rather well-understood subjects of condensed-matter physics, some of the most interesting fluids are formed by large chemically exotic molecules that pose a formidable array of challenges as diverse as chemistry itself. Polymer physics poses some extremely difficult questions in chain conformations, molecular dynamics, thermodynamics, and surface properties. Elegant techniques are required for the observation of dynamics on time scales from femtoseconds to hours. Massive computation is generally required for theoretical analysis of the disordered complexity of fluid behavior. Of great current interest are the liquid crystals formed by strongly anisotropic molecules that exhibit simultaneously some properties of fluids and solids.

Electrically conducting polymers appeal to the solid-state physicist as a potential source of new or enhanced phenomena, including superconductivity, charge transport, and nonlinear response. Collaborations between the solid-state physicists and synthesizing chemists have generated discoveries that present tantalizing prospects for future research in this field.

Geophysics

Geology and physics interact through the application of physics to understanding the structure and dynamics of the Earth and planets.

Geophysics has developed into an independent subfield of geology. In practice, the interdisciplinary interaction with physics is based more on conversion of individual physicists to geophysics and on adoption of physics-based methodologies than on the collaborative interdisciplinary research characteristic of the other interfaces described here. This characterization is, however, subject to some remarkable exceptions: geophysical fluid dynamics presents problems that test the fundamental understanding of turbulence and disordered nonlinear systems in general. Thus, there is intense interdisciplinary study in major geophysical institutions on a basic theory of turbulence. Some concepts in mathematical physics that offer the hope of more-powerful theoretical approaches to the complexity of geophysics are discussed in Chapters 6 and 7.

Finding methods to make accurate measurements of geologic variables challenges physics. The limited accessibility of the interior of the Earth, the depths of the seas and the atmosphere, as well as the sheer volume of data involved in characterizing geophysical phenomena call for heroic experimental measures. Measurement technology thus draws heavily on new physics. Space science has provided an important source of these new and powerful methods.

The most momentous development of the past few decades in geophysics is plate tectonics—the hypothesis that large plates on the Earth's surface move with respect to one another, cycling the ocean floor back into the mantle. Geophysical observations of alternating bands of ocean-floor rock magnetism were a key factor in establishing plate tectonics. The theory of plate tectonics is now being applied to the study of volcanism, mountain building, and mineral distribution through increased understanding of our planet's internal mechanics and the thermodynamic driving forces associated with internal temperature gradients.

The geophysics of the oceans and the atmosphere—the structure and dynamics of planets—all require sophisticated measurement technology. For example, basic seismography can now be applied to less than half of the Earth. Ocean-bottom seismology, only now becoming possible, will expand that fraction. Laboratory measurements at ultrahigh pressure of the properties of rocks and minerals today provide the primary basis for determining the equation of state of the Earth's interior.

Physical methods for mapping, long-baseline interferometry, seismology, gravity, terrestrial magnetism, sonic ocean probing, and atmospheric probes were early measurement devices. In recent years laser altimetry, precision radar, remote optical and physical sensing from Earth satellites and planetary probes, mechanical reflection probes, and electromagnetic probes of the Earth and oceans have been added to the repertoire. The overwhelming volume of data generated by such measurements calls for extremely large-scale data-processing and computing capability, including development of a global array of modern digital seismic instruments. Tomography, developed by physicists for construction of human organ images, is now proving valuable in depicting the interior of the Earth and oceans.

Geophysical research has important immediate applications to the environmental problems of society. Prediction of seismic activity, volcanoes, climatic changes, and weather are all important to the economy and welfare of mankind. The search for fossil fuels and the understanding of the environmental impact of energy production and consumption are based on geophysics. Some of these applications are outlined in Chapter 5, on geophysics; others are raised in Chapter 11, on applications of physics to energy and the environment.

Mathematical Physics and Computational Physics

From its beginnings, physics has relied on mathematics to formulate and express quantitatively the basic concepts of physics. Therefore, the limits of mathematical understanding present implicit limits to the development of physical understanding. Today physics itself has become so sophisticated that new discoveries often require deep understanding of both sciences, and synergistic interchanges of ideas in mathematics and theoretical physics enrich both fields. Mathematical physics is currently a lively arena for discoveries of immediate relevance to the understanding of complex observable phenomena, particularly through the development of understanding of nonlinear systems that may display disorder and chaos.

The hierarchy of the elementary particles is based on the development of nonlinear gauge field theories. Field theories have guided experiment to a long succession of discoveries of fundamental particles predicted earlier by theory. This success, perhaps, presages the unification of cosmology and elementary-particle physics. Physicists and mathematicians alike have contributed to these theoretical developments, and there is noticeable role switching, as new mathematics is introduced by the physicists and new physics by the mathematicians.

Statistical physics now begins to encompass chaos. The mathematics of nonlinear equations that display disorder and instabilities provides models for disordered physical systems, such as turbulent fluid flows, glasses, and ensembles of living cells. These mathematical developments portend new understanding of nature's most complex disorder.

This lively development in mathematical physics is abetted by the development of powerful, convenient computational machinery and concepts. The complexity of the problems addressed in modern mathematical physics frequently calls for numerical analysis. Today new theory can be created through computer analysis. Nonlinear problems with many degrees of freedom are now so prevalent that computational physics has become an indispensable aspect of theoretical physics. University-based centers for nonlinear studies have played an important role in the development of this field.

Computers have now become a means for generating new theoretical discoveries, transcending their more traditional roles as mere tools for measurement and analysis. Nonlinear equations in nuclear reaction theory, chemical kinetics, plasma simulation, galactic dynamics, and quantum field theory are some of the outstanding successes in computational physics. Elementary-particle physics, quantum field theory, and renormalization theory have all assimilated great amounts of computational physics.

In statistical physics and condensed-matter physics, the only possible solution of many basic problems is through computational methods. The new mathematics being developed to deal with chaos, the disorder of fluid turbulence, and nonlinear dynamics is based firmly on interactive machine computation as the theoretical research tool of choice. Computational physics already constitutes a ubiquitous subfield of physics.

Virtually every volume of this Physics Survey emphasizes the need for larger or more interactive computer facilities. Computational physics—not just computers—is assuming a role in nearly every realm of physics. The diversity and speed of developments in computational methods support the growth of a computational physics that requires intense constructive interaction between physicists and computer scientists. The interaction should develop as true intellectual exchange rather than simply as an application of computer science to physics. The recently announced National Science Foundation supercomputer initiative will substantially improve university involvement and leadership in research on and with computers. It is welcome indeed.

TODAY'S SCIENCE/TOMORROW'S TECHNOLOGY: THE PROCESS OF INNOVATION

The applications of physics to technology generate substantial benefits for the U.S. economy. They are currently most conspicuous in the high-technology, high-growth areas, such as the electronics and information industries; but applied physics is an essential albeit often inconspicuous ingredient of innovation throughout industry. The number of physicists engaged in applications of physics and in the associated technologies far exceeds those engaged in fundamental research and academic pursuits. Indeed, discoveries and applications of physics substantially support the national infrastructure for the technological innovation process.

The economic benefits of innovation—the creation of new wealth for society—require successful consummation of the entire *innovative process*: the conjunction of invention and implementation. Typically, several apparently unrelated discoveries are necessary before a new technology can be successfully launched. Historically, the sources and timing of these discoveries have been disparate and difficult to recognize at the time. Of the several hundred tiny puddles randomly formed during a rainstorm, imagine trying to predict precisely which ones will contribute to the successful formation of a given stream as the rainfall continues. Attempts to devise funding strategies or to manage research to optimize its economic reward are analogous to trying to guide rain into some puddles and not into others. The successful pattern is difficult enough to discern after the stream is flowing; it is impossible to ascertain in advance.

Even if economic return were the sole reason to support research, our nation would have no choice but to support the widest range of fundamental scientific inquiry in order to take advantage of the serendipity that couples research, technology, and the economy. Because the ultimate usefulness of any given discovery is so unpredictable, the most appropriate sponsors of fundamental research are those with the broadest technical needs, namely, the federal government and a few major industries. Consequently, small businesses and industries have not often participated directly in the initial stages of the innovation process. The resulting lack of science-technology interaction in small industrial organizations introduces a severe handicap in dealing with technology transfer, that is, the transfer of new scientific discovery to development and commercial implementation.

One might argue that commercial success—even in businesses using the most advanced technology—has not required direct participation in the discovery or research phase. Indeed, the recent example of Japan appears strongly supportive of this view. Caution is in order, however, about the continued validity of this argument for several reasons. First, the Japanese successes of recent years have been more directly due to excellence in management, manufacturing, and marketing than to leadership in science. Second, as new technologies become more

complex and science intensive, the sequential progression of technological effort from research to development to manufacturing to
marketing will be supplanted by increasing branching and complexity.
Our major international competitors, particularly Europe and Japan,
are consequently *increasing* their basic research efforts at a time
when—despite recent gains—our relative national effort in the United
States remains below that of 15 years ago. *Fundamental research
activities are conspicuous by their absence from many of the largest
industrial concerns employing physics-based technology. The absence
portends future problems.*

The symbiotic relationship between research and science on the one
hand and technology and economics on the other has not only created
economically significant new technologies but has also formed the base
for new generations of increasingly sophisticated research endeavors.
The selected applications of physics that follow here are chosen to
illustrate these themes.

Progress in the Applications of Physics: Microelectronics

The electronics industry illustrates a key principle in the exploitation
of advances in physics for the benefit of society. It is useful to
understand the distinction between creating wealth and merely making
money. Societal wealth is increased only when the totality of goods and
services produced grows both in an absolute sense and relative to the
population. This means that productivity must increase. The ability to
amplify the productivity of human beings rests largely on technology.
In the industrial age, machines replaced or supplemented human
muscle power. In the postindustrial or information age, devices that
control the machines as well as an effective networking of human
talents made possible through enhanced communications are providing
further increases in productivity.

From the invention of the transistor until recently, the trend in
electronics technology has been to enable the experts to perform the
tasks of their professions more efficiently and more accurately by
providing a better set of tools. The coming decades will be characterized by the growth of a *societal information network*. Information
sharing, together with increased sophistication of tools among technological experts, will reach a level whereby each expert may have access
to all the information and knowledge of every other expert. In a sense,
then, the power and productivity of each element in society will be
greatly amplified.

The societal network will extend well beyond the mere sharing and proliferation of knowledge and information to the evolution of a more complex, highly productive society. Clearly, those nations that lead in the implementation of technologies that support such a goal will benefit most.

In the field of electronics, we have seen nearly a thousandfold increase in information-processing power since 1970, while the cost per logical operation has decreased by a similar factor. Progress in the physics of materials processing has enabled industry to incorporate into a single fingernail-sized silicon chip the computing power that two decades ago would have required a room-sized computer. Entirely new kinds of devices have been conceived that use photons (light) instead of electrons for the handling, processing, and modification of information. Miniature lasers and photodetectors are already in large-scale manufacture based for the most part on discoveries of solid-state and materials physics that occurred within the past decade. Optoelectronic devices that combine electronic and optical functions for undreamed of processing power are already on the horizon. Physicists have learned how to speed the flow of electrons through solid materials to the point where they begin to approach the speed of light. The combination of this high electronics speed with the ability to make nearly atomic-scale structures using methods originating in high-vacuum and surface-physics research promises computational and communication speeds that are thousands of times greater than anything available today.

The physics of disordered or amorphous solids has not only provided deep understanding of this novel state of matter but also has already shown capabilities of very-high-speed switching. The exploitation of amorphous semiconductors in technology is just beginning and could form the basis for a novel energy technology. In the pursuit of fundamental understanding of electronic materials, physicists have achieved atomic-scale insights into defects, dopant atoms, and interfaces between different materials that will ultimately determine the performance of future devices.

In electronics technology, a multibillion-dollar industry has emerged over the past several years capable of exploiting rapidly such advances in fundamental understanding. This industrial infrastructure is one of the most valuable assets of the United States in international competition. Indeed, it has been largely responsible for the term "high-tech industry."

Related future technological applications may be based on other physical phenomena, such as superconductivity. Superconducting Josephson devices, for example, may form the basis for new super-computers. Current research in physics on novel quantum-mechanical behavior for superconducting amplifiers offers the prospect of enhanced signal detection, which will not only further other fields, such as astrophysical observation, but will also permit technology advances in both information processing and storage that extend significantly beyond current levels.

The advances in magnetic processes for storage and manipulation of information as magnetic bubbles or on magnetic tape or magnetic disks have fostered magnetic memories with densities approaching one-hundred-millon bits per square inch. Current research indicates possible improvements of at least a factor of 10.

Optical Technology

Optical information technologies have, perhaps, grown fastest of all technologies during the past decade. It was only in 1973 that the first continuous room-temperature semiconductor laser was operated. Today such lasers not only operate reliably with lifetimes in excess of 100 years, but they have been made smaller than a grain of sand. They can send signals exceeding a billion bits of information per second. To achieve these gains, the basic physics of the laser, which flowed largely from molecular and atomic physics and optics, had to be importantly supplemented in technology by the physics of materials and semiconductors.

Basic research into the optical properties of materials and the fundamental processes of the absorption, fluorescence, and scattering of light have all combined with the invention of the laser to make possible today's ultralong-distance optical-fiber communications systems. The ability to detect and refine impurity atoms to levels below parts per billion in the materials that ultimately will form glass fibers has its origins in the atomic physics and spectroscopy of the mid-twentieth century. The novel processes based on the understanding of physical chemistry and kinetics that use gas vapors to fabricate ultrapure light-guide fibers are powerful illustrations of the transfer from fundamental research to important technology within the span of less than a decade.

New ultrasensitive photodetectors are based on understanding of semiconductor physics as well as on the newly realized ability to manipulate materials at the level of atomic layers to produce entirely new kinds of materials and properties. New types of lasers have been used together with novel optical-fiber light guides to transmit information at several billion bits per second over distances exceeding 100 miles without the need for any signal regeneration or amplification.

This capability opens the possibility of developing low-cost, high-speed information systems that operate over distances typically encountered among islands in chains such as Hawaii, Indonesia, and Japan.

The ability to carry so much information over a single hair-thin glass fiber allows the transfer of thousands of times more information than is possible with existing telephone technology. This information, in the form of high-definition switched color video, high-fidelity sound, data for computers and ultimately for life-sized, three-dimensional images moving in real time, will undoubtedly transform the ways in which our society works and lives.

New prospects for the processing of information through a marriage of electronic and optical phenomena are already beginning to emerge. Magneto-optic, electro-optic, and acousto-optic devices are already finding specialty applications. Entirely new phenomena discovered by physicists within the past decade offer prospects for processors and memory devices with orders-of-magnitude greater density and speeds than those of even the fastest and most miniaturized electronic computers available today. Optical bistability makes possible the prospect of an all-optical or photonic computer. Physics has already identified the fundamental processes and the basic concepts required. Further research can confidently be predicted to bring many of these ideas to practical reality and probably to commercial use within the next decade.

Instrumentation

The advances of physics instrumentation have traditionally served two major purposes. There are those advances that make physics research more precise or more effective, and there are those (sometimes members of the same class) that find application primarily in other fields of science or technology or perhaps in society at large. The laser is a prime example of a device that fits into both categories.

Large particle accelerators continue to play a dominant role in high-energy physics, nuclear physics, and plasma physics. Increasingly, however, condensed-matter and low-energy physics make use of large machines and central facilities. As physics research has moved to higher energies, the preceding generations of particle accelerators have passed on to widespread application in other fields of science or in technology. Ion-beam accelerators are now routinely used in the semiconductor industry. Neutron sources and synchrotron light sources are increasingly used in the study of condensed matter. New spectroscopies, such as x-ray-absorption fine structure and ultraviolet photoemission, are already greatly increasing our understanding of the chemistry and physics of surfaces, the behavior of matter in two dimensions, and the details of energy-band structure in more conventional three-dimensional materials.

Electron accelerators are now being modified or even totally redesigned to produce new classes of radiation exemplified by the free-electron laser, which offers the possibility for high-intensity tunable radiation at wavelengths from the far-infrared into the extreme-ultraviolet or soft-x-ray regions. Ultrahigh vacuum processing will surely become commonplace in the production of future generations of microelectronic devices. It has already proved to be an extremely useful aid in understanding the formation, atom by atom, of new materials on existing solid surfaces.

Diagnostic tools such as the electron microscope and high-power x-ray sources have revolutionized our ability to characterize materials with unprecedented detail. The ability to distinguish the chemistry of only a few atoms buried deeply in virtually any material will be absolutely necessary for future technologies as devices become ever smaller. The recent invention of the vacuum tunneling microscope has made possible for the first time real-space resolution of surface features on an atomic scale.

The development of ultrashort laser pulses by physicists only 3 years ago permits examination of phenomena in real time only a few quadrillionths (10^{-15}) of a second in duration. It has opened entire new fields of phenomena in physics, chemistry, and biology to detailed examination never before possible.

Physics-based instrumentation of direct benefit outside the field has included the development of medical instrumentation, such as the various tomographies (nuclear magnetic resonance tomography being perhaps the most exciting); the widespread use of lasers in industry for manufacture, monitoring, and control as well as the use of holography in the reading and reproduction of images and even in the decoding of information on product pricing; and the commercialization of laser spectroscopy apparatus for routine characterization in biology and analytical chemistry.

A resurgent field of great economic significance is that of manufacturing science. It embraces automated assembly and robotics for manufacturing but covers as well computer-aided design and process control. Manufacturing or production science must combine sensing and monitoring hardware aspects with sophisticated software and computer diagnostics and control. In future manufacturing facilities, the versatility will reside in the hardware, which will permit a variety of measurements to be made and changes in configuration to be implemented. The specificity will reside in the software, so that the distinction between manufacturing one type of product or another may be ultimately under control of the software program.

The major trends within physics as far as instrumentation is concerned are the push toward higher energies and larger machines for the high-energy and nuclear physicists; increased use of medium-scale machines such as synchrotrons and neutron sources for condensed-matter physicists; and increased use of microprocessors in small-scale individual laboratories for the control, acquisition, and processing of experimental data. The ability to run experiments under computer control has permitted experiments that would have been impossible only a few years ago.

Unfortunately, many U.S. university physics laboratories today lack modern electronic and computing equipment. Physics faculty and students are often thus forced to build equipment that they could purchase were the modest necessary funds available. Although there is some pedagogical advantage in the construction of equipment, we believe that it is more productive to build on the state of the art rather than to reproduce it.

Energy and Environment

The relationship between physics and the production and use of energy is much more complex than the mere application of physics techniques or principles to energy technologies; it involves the complicated feedback of applications-related discoveries to the fundamental research challenges of physics itself.

Among the important contemporary examples is the relationship between condensed-matter physics and solar electrical power generation and storage. Semiconductor physics research has made possible achievements of 20 percent solar-to-electrical conversion efficiency in single crystals. On-site *solar generation* of electrical power in small-scale typically Third World applications could soon prove practical as the conversion efficiencies demonstrated in the laboratory are brought further along in the development phase. The use of electron, ion, and laser beams in materials processing may soon affect solar-cell technology significantly. The role of physics-based instrumentation for analysis of both the materials and the electronic and optical phenomena that govern their device performance continues to expand.

The promise of controlled *nuclear fusion* for the generation of energy involves multiple challenges to the application of physics. Progress to date has been substantially propelled by physics discoveries such as high-field superconducting magnets, laser- and particle-beam-driven inertial-confinement schemes, plasma/electromagnetic-wave interaction phenomena, and particle/solid interaction effects. The last-named phenomena are particularly important because the fundamental reaction in controlled nuclear fusion liberates highly energetic neutrons, which inevitably interact with the structural materials of the reactor. Such interactions may lead to embrittlement, loss of ductility, swelling, creep, and other destructive effects.

Whereas solar and nuclear fusion technologies may offer significant promise for our long-term energy needs, *combustion* is at present responsible for more than 90 percent of the power generated in the world. Research in physics and physical chemistry related to improved efficiency of combustion processes clearly offers significant near-term potential returns. For example, a 1 percent increase in automotive fuel efficiency would save approximately $1 billion a year in energy costs in the United States alone. Related problems such as air pollution, acid rain, and combustion-driven corrosion are of extreme economic and social importance, and efforts to solve them will continue to benefit from the techniques and discoveries of physics research.

All methods for energy production, transformation, or transmission have significant environmental implications. Environmental physics is concerned primarily with the mechanisms, rates, and pathways for the transport of matter and energy through the atmosphere, the oceans, and the entire ecosystem. Atmospheric science confronts the complex problems of weather patterns, thermal fluxes, and radiation balance as well as the generation and transport of particles and pollutants. Advances in mathematical techniques for modeling complex fluid flows and chemical reactions are beginning to provide quantitative descriptions of the atmosphere's dynamics. New analytical techniques, such as neutron activation analysis and remote laser spectroscopy, are making possible unprecedented precision in determining the compositions and distributions of pollutants.

The overall radiation balance of the Earth represents one of the most important questions for the future of life on this planet. Probably the most important single environmental factor associated with the production and use of energy is the atmospheric concentration of carbon dioxide. The burning of fossil fuels over the past century has increased that concentration from 280 to 335 parts per million. At the same time, particulates generated by coal burning, for example, have changed the Earth's reflectivity. The net effect of these and associated phenomena on the average Earth temperature is a continuing source of controversy and debate. Because the systems are so complex and inadequately understood and because the carbon dioxide problem is so important, existing long-range fundamentally based research activities should be strengthened, broadened, and placed in an international context to increase underlying knowledge necessary to develop prudent strategies regarding the use of fuels in the coming decades.

National Security

A strong scientific base is essential to the national security. As a consequence, U.S. Department of Defense research and development programs represent a substantial fraction of the total national research and development effort. Of all the disciplines, physics has provided an exceptional fraction of the novel developments both in defense systems and in the verification of arms-control agreements. The past decade has seen the development of sophisticated laser systems for target designation, underwater communication, isotope separation, and meteorological long-range sensing. It has seen the development of atomic clocks for precise navigation and secure communication, the development of high-speed electronics made possible by compound semiconductors, and the development of magnetic bubble technology for ultrarobust, high-density memories operating in harsh environments. Indeed, as a market for the highest-technology products, the national defense establishment serves an important role in fostering and financing the developments of many technologies whose initial commercial potential alone would not justify the involved expense. Military systems, largely because of the microprocessor and both optical and electronic communications technologies, will become more intelligent and more network oriented. New means must be found to reduce the detectability of our military systems. Moreover, as potential adversaries' capabilities for concealment of weapons evolve, we shall require more-effective detection systems. These must be based both on more-sophisticated signal-processing methods and on a deeper understanding of the interaction of the various optical and electronic probes with materials. The current quest for directed energy weapons, if it is ever to succeed, must explore imaginative and currently unrealized applications of physics principles.

Strategies of national security continue to press the bounds of technology and even of basic physical understanding. From military systems to arms-control verification, basic physical analysis becomes an increasingly necessary ingredient not only of technological advance but also of strategic decisions. The critical importance of national security demands that all sectors (military, government, industrial, and academic) cooperate to ensure that the basic store of physical knowledge be both vigorously expanded and intelligently applied. An increased portion of the military research and development budget should be firmly assigned to long-range fundamental research. Especially in need of revitalization is the link with the universities where future scientists are educated and where so many of the basic advances in physics originate.

Medical Applications

In the medical applications of physics, both diagnosis and treatment have been revolutionized during the past decade. The most spectacular development is computer-aided tomography, for which two physicists were recently awarded the Nobel Prize in medicine. This conceptual advance has revolutionized x-ray radiology and ultrasonic scanning to diagnose disorders of the brain and blood circulation as well as malignancy. Computer-aided tomography techniques are also applicable to more-exotic physical measurements, such as positron-emission tomography and the promising technique of magnetic resonance imaging, which is just now becoming commercially feasible. The latter has been made possible not only by the technique of nuclear magnetic resonance coupled with computers but also by the availability of high-field superconducting magnets, both of which are based on physics research.

Lasers have become an increasingly important tool for surgery, making possible the effective treatment of disorders that formerly led inevitably to blindness. The carbon dioxide laser has been used in cancer surgery as well. Together with modern development in molecular biophysics, lasers can provide powerful analytical techniques for clinical biochemistry.

Within the next decade we foresee increased, accelerated applications of physical principles to the field of biosensors and real-time monitoring of crucial physiological parameters. The continuous accurate monitoring of blood pressure, blood sugar, heart rate, and other functions will become routine and inexpensive, thanks to the combination of sophisticated physical sensory techniques, miniaturized electronics and photonics, and low-cost microprocessors. All these technologies directly trace their origins to the fundamental physics of materials and phenomena developed within the past few years.

RECOMMENDATIONS

It is demonstrated throughout this multivolume survey that fundamental research in physics generates discoveries and understanding of substantial benefit to mankind. In this volume, the proposition is extended to fundamental research at the interfaces of physics with other sciences and shown to be similarly productive; furthermore, the essential role in technology of applications of physics depends on continuing strength in basic physics. The recommendations presented are intended to improve the effectiveness of physics research at its scientific interfaces and in its applications.

The implied categorizations into applications and interfaces are diffuse; many exciting discoveries transcend the boundaries of scientific disciplines, and distinctions between basic and applied physics are often solely in the eye of the observer. There is a continuum of activity in physics ranging from fundamental research to engineering applications and industrial product development that is characterized by complex interconnections, parallel and competing pathways, unpredictable time scales, and inescapable redundancy.

The unpredictability and complexity of this continuum preclude simple strategies for program optimization in fundamental research and limit the efficiency with which applied research can be directed to optimize its economic benefits. Because the applicability of scientific discovery is unpredictable, it is proper that purely scientific issues should dictate the selection of fundamental research areas. Our recommendations thus reflect two features of the innovation process—its unpredictability and its technological span from basic to applied. The recommendations also recognize the quickening pace and increasing complexity of the process, and their object is the long-term health of both the U.S. economy and the U.S. scientific enterprise.

The intellectual boundary between engineering and physics is vanishing in many areas of advanced technology, and the continuum thus created speeds technology transfer and innovation. The postgraduate career demands of most scientists and of engineers caught up in research and development pull them inexorably into this continuum and away from disciplinary boundaries. Of course, academic disciplines serve a traditional and necessary role in maintaining standards and systematic programs in education; although the best programs adapt smoothly to changing demands, the time may have come to initiate a strategic reconsideration of the formal disciplines to reflect modern science and technology. The engineering-physics and applied-physics programs in a few universities represent only limited response to a problem that appears overdue for a national study. Within the funding agencies, it may be appropriate to consider linking the corresponding parts of the basic science and engineering sections, and agency support of academic engineering research should cover the same long-range time horizon that agency support of basic physics research covers.

We recommend that universities and funding agencies organize to accommodate and enhance the engineering-physics interface in both education and research.

Although the proper role of government in industrial development of technology involves complex issues beyond the scope of this survey, several recommendations have surfaced in the deliberations of the Panel on Scientific Interfaces and Technological Applications. Participation of U.S. industry in basic physics research should, for the health of the economy, be extended beyond the current participating handful of large companies. The probability that results of sponsored basic research are likely to be applicable to sponsors' problems increases with the breadth of their interests. Thus investment in basic research is most appropriate for large, versatile sponsors. Nevertheless, small firms that are active in advanced technology benefit from participation and association in fundamental science through intellectual stimulation and technology transfer at the least. Facilitation of collaborative activities can be an effective approach to coupling smaller firms to fundamental research. Government programs to encourage industrial innovation, such as the Small Business Innovative Research program, are also laudable. However, the funding algorithm for this program in particular has the effect of subsidizing private commercial development. Care should be taken to ensure that such subsidy does not occur at the expense of long-range university research.

Tax policy should encourage both in-house industrial research and cooperative research programs with universities and government laboratories. Antitrust laws and policies should be modified or reinterpreted to facilitate and encourage industrial and institutional collaborative research combinations and associations that enhance the national competitive position and economic welfare.

The interfaces of physics with other sciences call for special attention by the universities, the scientific societies, and the government agencies to accommodate their interdisciplinary essence. Appraisal of research proposals along the lines of the traditional scientific disciplines helps to preserve standards of quality, but greater flexibility and innovation in the funding agencies is needed to accommodate the intellectual activity at the interdisciplinary interfaces.

Funding agencies should devise procedures to evaluate and support interfacial and interdisciplinary research collaborations involving participants from deep within the associated disciplines. Special provisions should be made for initial research grants for young faculty beginning interdisciplinary programs.

A notable example of an administrative scheme that created the successful nucleus of a productive interaction and was a strong factor in the actual founding of the field of materials science is the set of interdisciplinary centers that has matured into the materials-research laboratories found on several campuses today. Biophysics and medical physics are interdisciplinary fields that would benefit from similar centers.

Universities should encourage the formation of interdepartmental, interdisciplinary research and research centers to attract and utilize interdisciplinary research funding to transcend communication barriers among the traditional disciplines and provide multidisciplinary and transdisciplinary education.

We note that the large-scale applied research and development programs of the mission-oriented agencies, in funding the federal government's research and development programs, place heavy demands on the scientific community. Consequently, it is appropriate that those agencies should share both in the support of basic research in related fundamental areas and in the education of scientists to contribute in these areas. Two such areas are addressed specifically in this Physics Survey: plasma physics, which is strongly supported with the specific goal of developing controlled nuclear fusion power, and national security.

It is our recommendation that the U.S. Department of Defense restore its investment in long-range fundamental research to pre-Mansfield Amendment (1970) levels in order to enhance the Department's connections with the physics research community for the mutual benefit of science and national security. The general fields of research selected for supplementary funding should be chosen to meet needs for applicable physics in mission-oriented agencies.

We recommend that mission-oriented agencies support fundamental physics research at levels appropriate to sustain the research and development program of their missions. With guidance from the research community, these programs should adopt long-range goals and a broad selection of topics, and they should incorporate peer review for quality control.

2

Biological Physics

INTRODUCTION

The magnificent complexity of life that we study as biology reflects ultimately the underlying principles of physics. The goal of biological physics is understanding of this physical basis of biology along all the complex pathways from atomic and nuclear physics, quantum mechanics and statistical physics, through the biopolymers (proteins and genes) and supramolecular structures to individual cells, and finally to the behavior of organisms. Biological physics comprises both applications of physics and fundamental problems in physics at the interface between physics and biology.

Biophysics serves humankind through its part in all applied biological sciences and through medical physics. (Medical physics is discussed in Chapter 13 as an application of physics, whereas biological physics is reported as an interscience interface because the present scientific trends and orientation of these two related fields are quite different.) Today's biophysics anticipates tomorrow's medical physics with a trend from large-scale diagnostic imaging to probes of cellular and molecular-scale processes.

New developments in both physics and biological science portend extraordinary opportunity in biophysics: Physics has developed to the stage from which it can broadly address the complexity of biological systems from basic principles. Physical theories of the hierarchy of

26

disordered systems begin to encompass biological systems' variability, and computational power begins to bring proteins and genes within reach of first-principles calculations. Proteins and DNA present great fundamental challenges as complex physical systems. Experimental sensitivities of physical probes now reach to the level of sparse molecular populations of individual cells and probe the correlations of complex cellular networks; neuroscience and cell physiology pose fundamental problems of biophysical organization that are beginning to become approachable through modern physical theory of complex systems.

Because multidisciplinary collaborations drawing on diverse specialties from cell biology to mathematical physics are frequently advantageous in biophysical research, we recommend organization of research funding to accommodate and promote it.

The new biotechnology provides through gene manipulation the tools for genetic engineering. Cloning of rare gene products makes available large quantities of protein for fundamental studies. Protein engineering through directed manipulation of DNA sequences permits systematic modifications of natural enzyme structures. Stimulation or suppression of the expression of selected genes in living cells and organisms permits direct observations of function. Monoclonal antibodies provide molecular specificity for biophysical probes in living cells. Physical studies of protein configurations modified by genetic manipulation offer vast potential to develop the basic understanding of protein structure and function needed for effective protein engineering capability.

The molecular understanding of photosynthesis is only the beginning of applications in agriculture; biophysical studies of climatic tolerance and injury in plant cells and the use of physical probes in plant-cell physiology are promising trends. In the pharmaceutical industry sensitive physical assays have become essential, and the potential of protein engineering is motivating fundamental research.

Advances in physical measurements have generated many important discoveries in biology, frequently following rapidly after new developments in physics. Not only have sensitive measurement technologies permitted the study of minuscule quantities of material at high precision but also the strategies of physical theory have provided penetrating views of biological phenomena. Many physical techniques have become so fully integrated into biological research that their origin in physics is forgotten until some underlying physical advance in the method provides a reminder; recent examples include various spectroscopies, electron microscopy, x-ray crystallography, and nuclear resonance. The synergistic combination of biological physics and biotechnology, especially genetic engineering, lends new power to biophysical research.

Three levels of biological problems stand out in current molecular biophysical research: (1) molecular biology—proteins, DNA, and RNA; (2) membrane and cell biology—cellular organization and function; and (3) cooperative multicellular systems such as the brain. A broad range of experimental and theoretical physics is involved in addressing these problems. Five themes of biophysical research are selected to illustrate extraordinary recent progress and current opportunities: (1) molecular physics of biopolymers, (2) experimental methods of molecular biophysics, (3) biophysics of membranes and cell physiology, (4) biophysics of brain and nerve, and (5) theoretical biology. As is typical of biophysical problems, our themes are interwoven and interdependent.

The *biopolymers*—the proteins and polynucleotides—are the macromolecules, with the lipids and polysaccharides, on which the chemical machinery of life is based. Vast and diverse knowledge of their physical and chemical structure, dynamics, and function has accumulated. Now the goal is to extract the basic principles of protein structure and function from an enormous variety of complementary experiments and empirical models. Genetic engineering extends the realm of protein biophysics to elucidation of protein structure and the creation of novel proteins.

The *physical methods* of molecular biophysics are discussed in the second section of this chapter. They represent a large fraction of recent biophysical research. A continuing stream of physical technologies has become absorbed into biology. Application of x-ray physics, advances in magnetic resonance and laser techniques, optical microscopy, and molecular spectroscopy have all become established as powerful biophysical tools.

Biophysical research on *membranes and cell physiology* has focused on the biological membrane processes that mediate all communication between the cell and its environment. Cell physiology involves complex molecular function and supramolecular organization of structure and dynamics of cell membranes and organelles—sometimes only a few thousand molecules per cell provide total control. The processes of visual transduction, auditory transduction, nerve signal transmission, ion channel molecules, ion pumps, and the bioenergetic machinery in the cell membrane are all exciting areas of research. Membrane fusion, a hydrodynamic process involved in excretion, hormone control, and signal transmission, has become a focus of research that requires more basic physical and physiological facts.

The biophysics of the *brain and nerve* amplifies many features of cell membrane physiology. Essential are the transmembrane channel molecules responsible for electrical signals. Current research aims to understand mechanisms of ion selectivity and gating at the molecular level. Understanding the global organization of the multicellular assembly of the brain (how memory works, how information is processed and perceived) is a major challenge for biophysical research. New concepts in theoretical physics and mathematics appear to be applicable to organization of neural networks. There is also a lively reciprocal flow of insights from studies of the brain to the development of concepts of artificial intelligence, computational strategy, and robotics.

Theoretical biology is generally becoming an important component of biophysical research at the levels of basic molecular structure and dynamics, cellular functions, and cooperative cellular systems. Much new mathematics and recent physical theory are proving highly effective tools for understanding the complex hierarchy and partial disorder of biological systems. In recent years theorists have become closely associated with experiment in effective interdisciplinary synergism. Thus, the span of physical disciplines involved in research on biological problems now extends from the most abstract mathematical theories of chaos, automata, and nonlinear dynamics to the most exacting experimental technologies.

BIOLOGICAL MACROMOLECULES

Biological macromolecules have evolved to perform specific functions. To follow the same selection principles in genetic engineering, we must understand the function of biological macromolecules on the basis of their chemical and geometrical structure. Both the polynucleotides, which carry the genetic information, and the proteins, which make up the molecular machinery of the cell, are linear polymers of covalently linked building blocks that differ only in their distinctive side groups. Their characteristic three-dimensional structures are dynamically stabilized by numerous weak interactions. In the polynucleotides, sets of four bases with matching hydrogen bonds form complementary pairs that play the fundamental role in structure and function. In DNA the resulting polymer usually assumes the geometry of the famous Watson-Crick double helix, a highly periodic molecular structure composed of many millions of atoms. Other less regular nucleic acids form important but less permanent parts of the cell's machinery. Recent discoveries of new DNA helix structures have redirected attention to the basic physics of polynucleotides; the molecular mechanisms of interaction between DNA and the regulatory histone proteins are being uncovered.

Proteins are linear polymers each containing hundreds of amino acids. The 20 different side groups permit an immense number of conceivable sequences, but the naturally occurring proteins represent only a small subset of the possibilities, each with three-dimensional equilibrium structural distributions controlled by weak interactions. The prediction of the equilibrium structure of a protein, given the amino acid sequence, remains an unsolved problem, although current empirical approaches have generated some success. Many proteins can be reversibly unfolded and refolded within seconds, suggesting a predetermined kinetic pathway rather than a random sampling of all conceivable configurations. The elucidation of the mechanism of this dynamic process is of fundamental interest.

Biological macromolecules are distinguished from the standard solids or liquids of condensed-matter physics by their size, by their linear polymer structures, and by their distinctive interatomic forces. Each macromolecule contains typically more than 10^4 atoms; its structure accommodates large, cooperative conformational fluctuations that provide for essential functions. For example, enzyme function is based on the binding of substrate molecules at specific binding sites, often with induced conformational changes involved in cooperative chemical reaction or catalysis. This is shown for a typical enzyme in Figure 2.1. The unique system of secondary forces and linear structure permits the large fluctuations associated with these functions.

Since the biopolymer backbone consists of relatively strong covalent bonds, whereas the interactions between side groups are quite weak, the normal vibrational mode frequencies cover the broad range from 10^{11} to 10^{14} Hz. In principle, all these modes and the slower thermally activated transitions could be functionally important in chemical kinetics, but fortunately some reactions can be approximated at low temperatures with just a few relative motions. There is hope that consistent connections between fluctuation dynamics and chemical reaction kinetics can be found. These unusual fluctuation dynamics, that is, the large amplitudes and the broad range of characteristic frequencies, are being studied by computer simulations of molecular dynamics based on theoretical concepts developed to understand the physics of fluids. At present, small proteins can be studied empirically with encouraging results. The future is promising as more-powerful

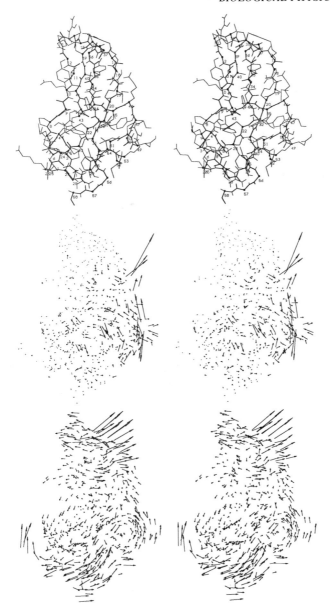

FIGURE 2.1 Molecular dynamics. Enzyme function depends on dynamical fluctuations of protein molecular conformation. Here normal-mode calculations plotted in
stereo show displacements associated with two of the normal modes in the enzyme
bovine pancreatic trypsin inhibitor. SOURCE: N. Go, T. Noguti, and T. Nishikawa,
Proc. Natl. Acad. Sci. USA 80, 3696 (1983).

computers make possible the more nearly accurate calculations that should provide incisive new insights. Development and application of fundamental theory of molecular physics to the biopolymers is an approach that shows great promise and should be supported with the necessary provision of large-scale computing power.

The functional transport of ions and electrons through proteins raises some particularly profound problems. In many essential enzymes, such as those that compose the reaction center described below, it appears necessary for physical electron transport by tunneling to occur before essential charge movement can take place, but basic details remain controversial. Similarly, the mechanisms of ion transfer across membranes induced by the membrane ATPases that serve as ion pumps seem to defy definitive determination. We shall return to another difficult problem—ion selectivity in ion channels—in discussing biophysical problems in neuroscience.

EXPERIMENTAL METHODS OF MOLECULAR BIOPHYSICS

Notable advances during the past decade have derived from x-ray and neutron crystallography, electron microscopy and diffraction, magnetic resonance (both nuclear magnetic resonance and electron paramagnetic resonance), optical absorption spectroscopy at wavelengths from infrared to x rays, laser Raman spectroscopy, fluorescence spectroscopy, and many physical probes of chemical kinetics. New capabilities for physical measurements of structures and processes provide much of our progress in biophysics.

X-ray crystallography has shaped our view of nucleic acids and proteins most profoundly and is likely to remain a rich source of information. For biopolymers the major breakthroughs came only in the 1950s with the discovery of the DNA double helix by Watson and Crick and the structure determination by Kendrew and Perutz of the oxygen-carrier proteins myoglobin and hemoglobin. Today's powerful x-ray sources and computers permit elucidation of the structure of very large proteins at high resolution; for example, catalase is an enzyme that measures roughly 10 nm × 10 nm × 5 nm and contains four identical subunits of 506 amino acids each. All amino acids have been assigned unambiguously at a resolution of 0.25 nm, explaining many interesting biochemical details. Figure 2.2 shows an x-ray diffraction pattern and a computer-simulated diffraction pattern for a protein crystal.

Synchrotron radiation offers a powerful source of high-intensity x rays for experiments. In the study of time-resolved diffraction of

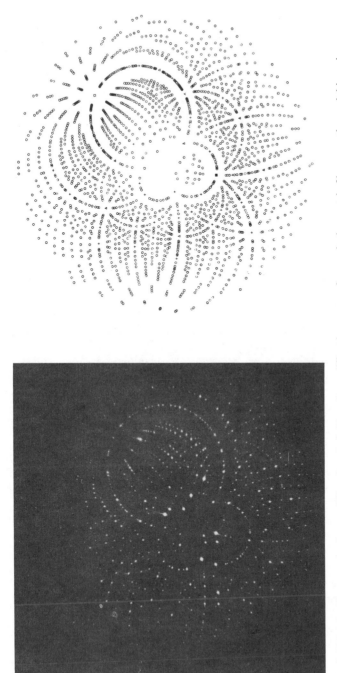

FIGURE 2.2   Time-resolved x-ray diffraction. *Left*: X-ray Laue diffraction pattern from a single crystal of horse methemoglobin, using x rays derived from the Cornell High Energy Synchrotron Source with a mean wavelength of 1.30 Å and a bandpass $\Delta\lambda/\lambda$ of 0.30. General crystal orientation; 1-min exposure on Polaroid Type 57 film. SOURCE: K. Moffat, D. Szebenyi, and D. Bilderback, *Science* **223**, 1423 (1984). Copyright 1984 by the AAAS. *Right*: Corresponding computer simulation.

contracting muscle fibers, for instance, the thousandfold increase in intensity over conventional sources has provided dynamical data on conformational changes during contraction. Dynamical x-ray diffraction and absorption fine structure experiments using synchrotron radiation are also being applied to enzyme kinetics and membrane phase transitions. Functionally important transient structures that emerge as intermediates in biochemical reactions are being studied by time-resolved x-ray diffraction and by extended x-ray absorption fine-structure (EXAFS) spectroscopy. The time stability of intermediates can sometimes be extended by optical pumping, rapid flow transfers after excitation, and low-temperature trapping. As synchroton radiation sources become more available, they can also enhance standard protein structure determinations as well.

Single-crystal neutron diffraction also provides a powerful probe, but its use has been limited by availability of suitable neutron beams. Inelastic neutron scattering, which probes the vibrational spectra of biopolymers, should be useful for protein and DNA dynamics as more cold neutron beams become available.

Genetic engineering is revolutionizing the production of proteins: the fastest way to obtain amino acid sequences now is by sequencing of the genes encoding the protein. Growing suitable crystals is still difficult, however. (Progress in understanding protein crystal growth would be invaluable; even the modest success already achieved with membrane proteins has resulted in biotechnological uses.) But once a crystal is formed, the chance that a structure can be determined is high. It is instructive to trace a few structural developments, since they illustrate the evolution of new concepts and the impact of improved experimental methods.

Myoglobin and hemoglobin serve an essential role as carriers of oxygen in tissue and blood. While the three-dimensional structures of these proteins could be rationalized on the basis of the known interatomic forces, the early models had problems accounting for the kinetics of their crucial function, the reversible binding of oxygen by the heme group buried inside the protein. Present views of protein structure and function resolve this apparent discrepancy by taking explicit account of structural fluctuations that give access to the buried binding site. Hemoglobin, the primary oxygen carrier of the blood, consists of four myoglobinlike subunits that bind oxygen cooperatively rather than as four independent subunits. Cooperativity is a widespread phenomenon in multisubunit systems, but the underlying mechanism continues to be elusive even in hemoglobin, an ideal model system. The important transient states in these molecules are illuminated by many time-resolved spectroscopies, including EXAFS and Raman spectroscopy. X-ray diffraction crystallography data, in principle, yield measurements of both static and dynamic atomic displacement from the most probable lattice sites through their effect on diffraction peak intensities expressed as a Debye-Waller factor and through diffuse scattering. The Mössbauer effect also provides a measure of the vibrational dynamics of some of the heavy atoms. Both of these methods have illuminated studies of the heme proteins and have shown potential for broader applicability.

The early model of DNA was so successful that the recent discovery of additional structural types of double helices came as a surprise. An equally exciting development concerns the regulatory proteins, which bind to segments of double helix with specific base sequences, thereby regulating the transcription of DNA. Restriction enzymes similarly have the ability to recognize certain DNA base sequences. Thus the question arises of how the apparently regular DNA double helix reveals its base sequence to a protein. X-ray diffraction has shown the dependence of specificity of enzyme-substrate binding on the matching of shapes of the molecules that allow favorable contacts to be formed. Recent high-resolution x-ray studies on fragments of DNA with different base sequences are said to demonstrate irregularities of the double helix. Raman spectroscopy, low-energy optical absorption, and new theoretical analysis of nonlinear molecular excitation modes are already providing an attractive basic physical description. Apparently, specific base sequences give rise to local distortions, which may be recognized by the regulatory proteins. Complementary work on restriction enzymes including the first crystal structure of a restriction enzyme complex have recently been reported. We are thus beginning to understand on an atomic scale some of the basic genetic control processes that are utilized at the molecular level.

Membrane proteins have essential functions in energy conversion, active transport, and the communication of the cell with its environment. Many are complex assemblies of subunits that perform a sequence of coupled reactions. Among the first to be studied and understood biophysically is the reaction center from photosynthetic bacteria, which converts light into electrochemical energy. Photosynthesis as the origin of our food chain is a broad research area that touches on many different sciences from physics to agronomy. The primary events of photosynthesis, light-induced charge separation and the subsequent chemical reactions, are becoming understood at the molecular level. Besides its scientific interest the photosynthetic reaction center may also have technical implications as the basis for efficient synthetic energy converters. A significant triumph of x-ray crystallography is the determination of the structure of the reaction center, which showed atomic placements that implicate mandatory tunneling processes in its function. The simplest reaction center consists of three proteins, two molecules of chlorophyll (which absorb the incident light), four chlorophyll-like molecules, and a ferroquinone complex. Years of research with a combination of different techniques have identified the sequence of events triggered by the absorption of light. The first electron is thought to transfer within picoseconds, followed by a second transfer to the ferroquinone. Quantum-mechanical electron tunneling is clearly involved, but it is still not clear how the rates are controlled by the electronic properties and relative positions of the active groups. A combination of electron paramagnetic resonance (EPR) with picosecond laser pulse spectroscopy has delineated the elementary processes from photoexcitation to electron transfer and radical formation and has given preliminary insight into the mechanisms of oxygen evolution in the green plants. Recent advances in the physics of lasers have increased the time resolution of pulsed laser spectroscopy another factor of 10, thus providing access to even faster biomolecular processes.

Most experimental methods in molecular biophysics are based on some type of spectroscopy. They often probe the electronic state or the dynamical properties of the complex active center of a biomolecule and have been particularly effective in the study of photoreceptors and the heavy-metal-based enzymes of bioenergetics. Useful wavelengths range from radio to x rays. We report here on a few recent new directions.

The profile of metal-atom EXAFS contains information about the electronic state of the metal and its surroundings. It is used to study metalloproteins and became practical with the introduction of synchrotron radiation only a few years ago. Although theoretical problems of interpretation remain, the full potential of the method should be realized as synchrotron radiation facilities become more available. Already EXAFS has provided insight into the binuclear functional groups that serve for oxygen transport and reduction in hemerythrin, hemocyanin, and cytochrome, for example.

Magnetic resonance has had a great impact on the biomolecular sciences since it was first discovered. Technological advances have kept the field productive, and now the enhanced resolution of the high-frequency and multiple pulse methods have made it possible to assign all the proton resonances in small proteins and nucleic acids, compare structures in solution with the crystalline structure, and probe dynamical features. Relevant isotopes other than ^1H, such as ^2D, ^{13}C, and ^{31}P, are now accessible with high resolution. Nuclear magnetic resonance (NMR) of phosphorus allows one to follow the transformations of the phosphorylated compounds that play a fundamental role in cellular intermediary metabolism. NMR now provides real-time measurements of the concentrations of many important compounds of metabolic control that make possible physiological studies in living animal models and even in humans. The high resolution provided by higher frequencies continues to find new applications, and pulse techniques developed for solid-state physics are applicable to slowly tumbling or nearly rigid biomolecules and membranes. The great promise of NMR imaging in medical physics is reported in Chapter 13. It may eventually be possible to incorporate this chemical sensitivity into a clinical imaging capability.

Many biological processes revolve around the transfer of electrons, most obviously in photosynthesis and in respiration, the source of energy of all organisms, but also in various enzymatic reactions. In all these processes compounds such as metalloproteins and free radicals with unpaired electrons can be studied by EPR. Multiple resonance or pulse techniques provide even more powerful tools as the technology transfers from physics research. EPR has contributed to elucidating the reaction mechanism of metalloproteins and the chemistry of radical reactions. It also probes molecular motion, and spin labels introduced into membranes provide a measure of the degree of order and the fluidity of the lipid bilayer that complements NMR data. An unexpected, conceptually important discovery came recently from a careful spin-relaxation study of metalloproteins, which showed that interactions can be described by fractals, surprising geometries with non-integer dimensions. The concept of the fractal description of irregular geometrical structures from coastlines to polymer clusters has recently attracted much attention in physics and is potentially applicable to many aspects of disordered biological structures.

Fluorescence spectroscopy provides dynamical information since emission probes the processes that occur during the fluorophore-excited state through effects on the quantum yield, the emission delay, and the changes in energy and polarization of the fluorescence. The random reorientation of the excited complex owing to thermal motion, the relaxation of the atoms surrounding the complex, energy transfer, and collisions with molecules such as oxygen that quench the fluorescence are among a host of possibilities that can be studied. The accessible time range is ordinarily within the lifetime of the excited state, typically a nanosecond, but processes as slow as 10^{-3} s are accessible by triplet conversion. The latest developments in light sources and electronics have also pushed the time resolution into the subpicosecond range and have thus opened a new window on fast processes. Fluorescent markers, especially those conjugated to specific antibodies, are powerful tools for mapping molecular distributions in living cells. Their versatility is enhanced by modern monoclonal antibody techniques.

Infrared spectroscopy has become useful for studying naturally occurring biopolymers through extraction of difference spectra by using the new Fourier transform instruments. For instance, in the optical pigment rhodopsin, changes in vibrational modes are discernible when the protein-bound molecule retinal is excited; in myoglobin the difference spectrum similarly reveals the additional oxygen-oxygen stretching mode that occurs on the binding of oxygen. The advent of the laser, with its intense monochromatic beam, permitted Raman scattering to become a practical tool for studying biopolymers. Some scattering occurs at a slightly shifted energy corresponding to the vibrational frequencies of the target molecules, and the scattered spectrum therefore yields the normal modes. Raman scattering and infrared absorption give complementary information. Tuning the incident laser light to an electronic transition for resonance Raman scattering strongly enhances the scattered intensity and provides information about the electronic state of the complex. The mechanisms of photon absorption and detection by the visual pigment rhodopsin and bacteriorhodopsin, along with reaction dynamics in chlorophyll, the light-harvesting pigment in photosynthesis, and the heme group, have yielded a wealth of data through Raman spectroscopy. Recent applications of Raman spectroscopy to DNA probe the dynamics of the various helical forms of DNA and the transitions among them.

BIOPHYSICS OF MEMBRANES AND CELL PHYSIOLOGY

The cell membrane mediates all cell interaction with its environment and regulates molecular traffic into and out of the cell. It consists of an amphiphilic lipid bimolecular layer less than 10 nm thick that provides a two-dimensional liquid environment for the set of specialized membrane glycoproteins that provide pathways for molecular transport, transmit transmembrane signals, serve as sensitive receptors for hormone and immunological responses, and provide intercellular communication and cell recognition. An epithelial cell membrane is shown in Figure 2.3. Membranes are involved in cellular mechanics by the adhesion, fusion, and stabilization of unique surface supramolecular

FIGURE 2.3   Membrane structure. Cells' membranes can join to seal along a line of tight junction, which prevents fluid flow between the cells in an epithelium. This picture shows a structure for the tight junction deduced from electron microscopy on fast-frozen fractured membrane fragments. It consists of a line defect in the lipid bilayer membranes that is understandable in terms of physical defects seen in smectic liquid crystals. SOURCE: B. Kachar and T. S. Reese, *Nature 296*, 464 (1982).

structures, such as microvilli, coated pits, and synaptic junctions. Each cell membrane may contain a trillion molecules in total, yet many individual molecular species serve their purpose with populations of only a few thousand. This delicate membrane-based machinery of the cell requires remarkably sensitive biophysical measurement methods to reveal the underlying mechanisms. Several current problems are discussed here, and more appear in the next section of this chapter, on the biophysics of brain and nerve, where the dominant role of transmembrane channels is described.

Microscopy with visible light and with electrons provides an essential fundamental tool for research in cell physiology where topography is critical. Digital video image recorders and analyzers with diode photon-detector arrays and electronic image intensifiers, sometimes with laser illumination for dynamic analysis, enhance the contrast of light microscopy for the detection of particles far smaller than the wavelength of light and the fluorescence from small numbers of molecules—all in living cells. Three-dimensional reconstruction from two-dimensional optical and electron-microscopy images is facilitated by large-scale computation using array processors and new software extending that developed for commercial television and satellite image analysis. (More-powerful procedures developed for military applications are unfortunately not yet available for use in biophysical research.) The capability of electron microscopy, although it is generally still restricted to preserved structures, has advanced with scanning instruments for transmission and backscattering images and for microbeam excitation of elemental x-ray fluorescence to map chemical composition. Now Rutherford backscattering and spatial-imaging mass spectrometry further enhance the topographical microanalytical tools.

Biophysical methods have recently become a natural partner of cell biology. The sensitive optical detectors and intense fluorescence labels specifically directed with monoclonal antibodies permit mapping of the distribution and dynamics of sparse populations of receptors and sometimes individual molecular receptors on living cell surfaces. A fluorescence study of low-density lipoprotein particles on a cell surface is shown in Figure 2.4. Molecular structures have been determined only for the few receptors that naturally occur in larger quantity, as in a special organ, but genetic engineering techniques offer the possibility of providing sufficient quantities of many for molecular-structure studies. Fast freezing of preparations followed by fracture to provide replicable surfaces for electron microscopy provide high-resolution maps of the topography and supramolecular organization of membrane and cell structures. The basic dynamics of cell-membrane receptor responses have posed some difficult biophysical problems that have now yielded to theoretical analysis, providing new insight about mechanisms of the chemoreceptor function in hormone response, chemotaxis, and immunology. Modern concepts in statistical physics have been assimilated into theoretical biology for studies of multivalent ligand binding.

The conventional fluid-mosaic model of the cell membrane has the membrane glycoproteins immersed in a two-dimensional lipid liquid. Hydrodynamic theory for fluid membranes suggests weak dependence of diffusibility on molecular size, so any molecule should diffuse across a cell in less than a minute. Extensive physical measurements of molecular mobility using time-resolved fluorescence microphotometry for fluorescence photobleaching recovery have shown, however, that in cell membranes all protein diffusibilities are many orders of magnitude smaller than expected for free diffusion because of ubiquitous

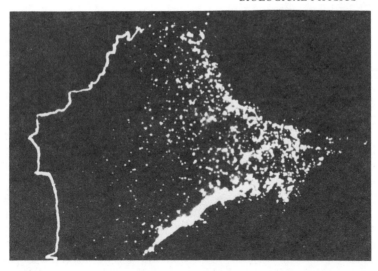

FIGURE 2.4 Cell biophysics. Individual receptor molecules on a cell surface are detected and mapped by fluorescence micrography using a brightly fluorescent ligand. About 75 percent of the light dots are individual low-density lipoprotein (LDL) particles bound to their receptors, and the remainder are clusters of two or three. These human mutant fibroblasts are derived from a familiar hypercholesterolemia in which, unlike normal fibroblasts, the few thousand LDL receptors on the mutant cells are not clustered in coated pits. In this case they have been segregated on the living cell surface by application of a small electric field in the direction of the arrow and are concentrated near one edge of the cell. The depleted edge of the cell is marked by a white line. SOURCE: D. W. Tank, W. J. Fredericks, L. S. Barak, and W. W. Webb. J. Cell Biol. *101*, 148 (1985).

natural restraints in the cell membrane. The mechanisms and regulation of molecular diffusion across cell membranes remain a current problem. Some natural constraints to stabilize surface organelles and systematic lateral transport are expected. Membrane flow and contractile processes generate systematic molecular motion that is involved in various essential cellular processes. The mechanisms of these important motions are under active study with the help of new physical methods that enhance sensitivity and image analysis in light microscopy. Fluorescence microscopy provides sensitive measures of membrane potential changes and intracellular pH and calcium-ion concentrations.

Fusion of cell membranes is a well-known and essential step in secretion, vesicle formation, viral infection, and endocytosis that presents a puzzling problem because it seems to be precluded by elementary physical considerations. Stable membranes repel each other at close separations because of the exponentially increasing hydration force required to displace water bound to the polar groups that stabilize the membrane surface. Current theoretical analysis of the physics suggests that nonlocal dielectric response dominates these interactions. At the biochemical level, it appears that the changes in membrane chemical composition that suppress hydration promote fusion: Proteins at some virus membrane surfaces may promote infection by exposing their nonpolar interiors so as to induce membrane fusion. Topological defects and cooperative distortions associated with the structural changes at phase transitions are also involved in fusion. The basic problem of membrane fusion has taken on practical importance in biotechnology through its role in gene and drug delivery systems. Membrane phenomena seem to be involved in effects of applied electric fields, which induce membrane fusion and move proteins on the cell surface. Medical reports have attributed a potential for the promotion of healing of injury to applied electric fields, although the mechanisms are not yet known.

Smectic liquid crystals provide useful membrane model systems through which membrane interactions, topological defects, and phase transitions can be studied. They resemble a stack of complete membranes separated by a thin layer of water. Modern physical theory of phase transitions is applicable. Molecular mobility on membranes can also be profoundly altered by phase transformations such as those in the liquid crystals. Membrane defects associated with phase transitions of charged phospholipids have been implicated in putative mechanisms of membrane fusion, osmolarity control, and enzyme regulation. Again, sensitive physical methods provide crucial information: fluorescence and NMR probe the dynamics and spatial segregation, synchrotron x-radiation diffraction is particularly useful for structural determinations, and electron microscopy of fractured fast-frozen sections delineate some macroscopic patterns. Lively research activity is beginning to generate some useful insights.

BIOPHYSICS OF BRAIN AND NERVE

The physical basis of the electrical activity responsible for signal generation and transmission in the nervous system has been recognized for 40 years, and many elegant details of this membrane function are understood. However, the biophysics of the brain continues to confound its students at many levels. Now several opportunities for progress in profound understanding of the most fundamental properties of the nervous system confront us. We discuss two here: membrane channels and multicellular organization of brain and memory.

New experimental technology provides direct access to the electrical activity of individual molecular channels in membranes, thus allowing the discovery and characterization of many of the channels present in the nervous system. Figure 2.5 shows sodium channel currents. Some of the molecular mechanisms of sensory transducers are partially understood: photon detection is understood at the molecular level, phonon detection is not; initial signal amplification in both cases remains elusive. Basic molecular mechanisms of memory and higher learning still elude us. But putative molecular mechanisms of learning have been identified in elementary organisms, and many important facets of the organization of optical image collection and projection on the visual cortex have been mapped in higher animals. These successes guide much current research. The system problems in organization of the brain, organization of memory, development of intelligence, and processing of sensory data remain profound. These problems have now developed a new dimension through connections with the technology of artificial intelligence, robots, and more-sophisticated computers. The current cross-disciplinary interplay of ideas among neurobiology, theoretical condensed-matter physics, and computer science is stimulating and lively. Prospects are good for the generation of profound new concepts.

Membrane Channels

The integral membrane proteins responsible for the electrical activity of the nervous system form transmembrane channels. There are perhaps several hundred channel species, but the precise number is not yet known. All channels permit the selective flow of ions through a pore of atomic dimensions. This flow is a selective, passive, possibly complex, diffusionlike process in which ions move down their electrochemical gradient. Different channel types have different selectivity; one may pass sodium ions but exclude potassium ions, whereas another may admit potassium ions but exclude sodium ions.

Most channels are gated; that is, their pores are not always open but rather are opened and closed through conformational changes of the channel protein. Two general classes of gating mechanism are recognized: voltage dependent and ligand dependent. The conformation states of voltage-gated channels are coupled to the transmembrane electric field through their dipole moments so that a change in the membrane voltage can change the probability that the channel is open

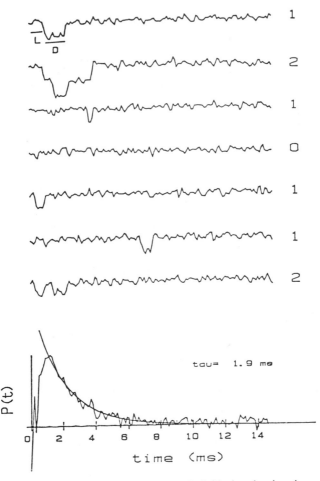

FIGURE 2.5 Biophysics of brain and nerve. Individual molecular channel currents recorded from a few sodium channels isolated in patches of cell membrane show the stochastic switching of the 1.2-pA channel current from which the opening delay latency L and open channel time D can be statistically derived. The opening probability P(t) for the sodium-channel component of the macroscopic-nerve action potential can be derived from the single-channel measurements and compared with actual nerve signals to identify the contribution of the sodium channels. SOURCE: R. W. Aldrich, D. P. Corey, and C. F. Stevens, *Nature 306*, 436 (1983).

or closed. Binding of specific small ligand molecules induces the conformational change that opens ligand-gated channels. Usually, a particular channel is either voltage or ligand gated, although sometimes it is both. Voltage-gated channels are responsible for the nerve impulse and for the encoding and transmission of information in the brain, whereas ligand-gated channels function in cell-to-cell communication at synapses.

Despite rapid advances in recent years, we still lack satisfactory theoretical explanations for gating mechanisms and pore selectivity. The central problems remain: How do channels gate and how do they discriminate? The main biophysical principles involved are understood, but techniques for relevant high-precision measurements are still developing. The recent development of single-channel recording methods for measuring the currents flowing through the pore of just one channel is providing much needed information. However, understanding of channel mechanisms requires not only functional but also structural analysis. Information on the primary structure of channels is coming from protein biochemistry and particularly from the sequencing of cDNA clones coding for the proteins involved. Higher-level structural information is derived from a theoretical analysis of primary structure and from diffraction experiments. Methods for obtaining three-dimensional crystals of channels are just now being developed and should yield high-resolution structural information. Structural analysis should become an important area of channel research. Theories can be tested by perturbing the structure of the proteins involved by using techniques of molecular biology such as site-directed mutagenesis. For example, one could formulate a theory about the role of a particular amino acid that forms part of the walls of the pore in discrimination among different ionic species and then predict the effects of changing it. The required structural modification could be made and the predictions of the theory tested.

The ligand-gated acetylcholine receptor (AChR) is at present best understood since the large quantities of receptors available from electric organs have facilitated biochemical studies. These AChRs appear electrophysiologically identical to receptors from the neuromuscular junction when reconstituted into lipid vesicles. All the subunits of the AChR have been cloned and are being reassembled, portending early progress in this case. By using a combination of biophysical, electrophysical, and molecular biological methods, we have the prospect of developing a detailed understanding of channel mechanisms and thereby an insight into the basis for the brain's electrical activity.

Organization of Brain and Memory

The cellular machinery of neurobiology displays some exquisite analytical tuning in signal propagation and processing. A highlight of neurobiology is the remarkable understanding of the organization of the visual cortex. Many aspects of the correspondence of its spatial organization to pattern recognition have been worked out. Dendritic nerve cells involve many branches with many intracellular connections at synapses; this system serves as an analog computer in signal processing and storage. Modifications of the strength of the synaptic connections are implicated in learning. But mechanisms for both fast temporary memory and long-term memory have yet to be found.

Discovering how the brain is organized as the most complex known computer in solving real-world problems in real time is a great challenge to theoretical biophysics. Our biophysical studies of the neural hardware and the wealth of available anatomical data leave the brain's organization, viewed as a wiring diagram, sketchy at best. Current experimental capability limits signal recording to, say, 10 or 20 cells in a network that may contain tens of thousands. New physical methods of optical recording of membrane potentials may help, but the experiments have far to go to match other areas of experimental physics. New experimental discoveries in anatomy, biochemistry, electrophysiology, and biophysics have a way of revising profoundly our views of brain function. One approach to theoretical formulation of the organization of the brain is to reduce the complexity by expressing it in terms of the simpler laws governing its component parts. Such reductionist extrapolations have yielded some understanding of many of the complex partially disordered phenomena of physics. Critical phenomena, dissipative systems, stellar atmospheres, and hydrodynamic chaos are surely unexpected guides to understanding the organization of the brain. The theoretical physicist hopes that the apparent complexity of the components and architecture of the brain is to some extent due to our present inability to recognize the simpler system of basic rules that generates the system. Ideas for such fundamental rules for neural networks are stimulated by the many recent successes in cooperative systems of condensed-matter physics. Analogies with disordered interacting arrays called spin glasses are particularly appealing: Application of their cooperative features in representing memory states reflect distinctive features of the human brain, including errors encountered when overloading, unlearning, and the chaotic states that are supposed to be associated with certain nervous disorders.

Neuroscience is presented here as the source of challenging fundamental problems for scientific exploration. Humankind, however, would look to neuroscience research to discover how and why the brain fails us and thereby support medical treatment. Medical physics has contributed new and powerful tools for macroscopic mapping of some physiological functions in the brain; notable among these are x-ray tomography, NMR scanning, magnetoencephalography, positron-emission tomography, and digital differential radiography. Neuropharmacology has revolutionized the treatment of many mental disorders, and one can hope that basic understanding of the molecular and organizational biophysics of the brain will serve in the next generation of medical advances.

THEORETICAL BIOPHYSICS

Physical and mathematical theory appears above as a component of each of the illustrative themes but is again addressed here to present productive current trends that we label theoretical biology. Population genetics, ecology, epidemiology, physiology of multicellular systems, and the mechanics of motility from the flight of birds to slime mold aggregation represent some of the theoretical problems traditionally addressed with success by theoretical biology. Here we emphasize a new trend in theoretical biology toward applications of microscopic theory and complex systems theory at the cellular and molecular levels. Revolutionary progress in understanding various processes in cell biology is under way through application of recent theoretical discoveries in mathematical physics and condensed-matter physics. The theoretical search for order in biology has often faltered in dealing with the inevitable partial disorder of biological systems. Now mathematical physics is beginning to deal successfully with disordered systems in condensed matter, for example, in critical phenomena, amorphous glasses, chaotic flows, and disordered dynamic patterns. These concepts appear to be broadly applicable to biological systems and are discussed in other chapters of this volume.

In developmental biology, ideas about how zebras get their stripes and how cellular slime molds aggregate are being based on theories of pattern formation in nonlinear continuum reaction-diffusion systems and in hydrodynamic instabilities. Systems theory concepts also provide for analysis of systems of cells. For example, discrete mechanical models of the system of cells in the developing embryo indicate that individual cells subject to chemically controlled cytoskeletal contraction appear to develop the standard embryonic patterning as a natural consequence of their growth. Geometric scaling theory, the concepts of fractal dimensionality in diffusion, and cluster aggregation all have biological applications not only to clustering of cells but in molecular reactions and even in description of the fluctuations of protein and DNA structure. Solitary waves (solitons) on the helical structure of DNA provide a representation of structural transitions that may suggest mechanisms of control of gene expression, a crucial basic problem of molecular genetics.

Like the human brain, the human immune system consists of a collection of about a trillion cells. Its function, to protect from disease, is regulated by complex interactions that provide global system operation and regulation, which are in turn subject to numerous disorders. As experimental definitions of the cellular and molecular ingredients of this system improve, the basis and the need for a comprehensive theory of this cooperative system are evident.

Applications of the mathematical theory of nonlinear dynamics are explaining many critical biological processes based on limit-cycle oscillations, particularly in biological clocks from the heart beat to the menstrual cycle. Even the daily rhythms of life may be attributable to oscillatory enzyme processes. Cardiac arrhythmias and fibrillation are understood as oscillator instabilities associated with the notion of reentrant disturbances in the mathematical system describing the coupled system of contracting heart muscle cells. The theoretical descriptions of the regulation of the heart beat are remarkably accurate and detailed. The physiological response to electrical stimuli is highly predictable and sufficiently reliable to form the basis for pacemaker diagnosis and design. Similar theory describes peristaltic motion in the gut and bursting irregularity of insulin secretion from pancreatic beta cells. In these areas the connections between experimental observations and the mathematical analysis have become well established in many laboratories.

The kinetics of receptor-ligand reactions on cell surfaces that mediate transmembrane signals or ligand internalization often involve multivalent ligands that induce clustering or trigger discontinuous switching signals. In the allergic response, cross-linking of IgE antibodies on mast cells may either trigger histamine release or alternatively desensitize the cells. Theoretical analysis of the basis of these signals is providing new insight into allergic response. It may seem strange to analyze hay fever in terms of the theoretical instabilities of a multidimensional lattice, but in fact lattice gas models are highly applicable to allergen sensitivity. Analogous kinetic problems arise in immunology, chemotaxis, and mitosis. Modern physical experiments are providing data to test theoretical ideas about these crucial processes in cell biology, which is advancing rapidly on the basis of lively interdisciplinary exchange.

Molecular biophysics has its own theoretical component, which has already been discussed. Here we have outlined a new trend in conjoined theoretical biology and theoretical biophysics in which modern concepts of statistical physics and condensed-matter physics are applied to biological systems at the cellular level. The cooperative dynamic processes at the level of molecular reaction and transport kinetics, cellular organization, and cell system organization are proving particularly amenable to these approaches. In many cases the theoretical analysis makes possible systematic experimental planning of the characterization of seemingly chaotic problem systems. We anticipate that theoretical biophysics will become a routine ingredient of cell biology, but maintenance of interdisciplinary connections with the applied mathematics and theoretical physics communities appears necessary to accommodate the continuous flow of new and relevant concepts from mathematical physics.

CONCLUSIONS AND RECOMMENDATIONS

We have defined biological physics as the application of physics to biological problems and as the scientific interface between biology and the underlying physical principles. The complexity of biological problems at all levels and the sophistication of the applicable physics of measurement technology and the theoretical concepts raises an important practical issue for this field: How is it possible to acquire the range of expertise needed for effective research at this exciting frontier? Often collaborations provide an effective pathway for researchers, but the diversity and sophistication of the specialties that are becoming involved compound the difficulties.

Today's biophysicist may have been trained in any science. Many begin with a Ph.D. degree in physics or physical chemistry or with an M.D. degree, since there are relatively few formal graduate school programs in biophysics, and have acquired multidisciplinary training through experience or multiple degrees.

The graduate student in biophysics is hard pressed to learn enough modern physics and enough biological science; it may be appropriate to recognize the extra burden through multiple degree programs. The success of M.D.-Ph.D. programs in educating biophysics researchers suggests the potential of joint programs between physics and other areas of basic or applied biology. With the growth of basic research in plant science and biotechnology, multidisciplinary training programs with physics and these areas of applied biology could be highly effective.

Few academic physics departments now include biological physics as an option within the discipline of physics. Our experience suggests that the physicist finds it easier to learn the necessary biology for biophysics research than the biologist does in trying to learn physics. It is timely to recognize that the rich and uniquely significant dynamic physical systems presented by life and its origins are intimately associated with the rest of physics, including particularly germane problems in the current study of the general properties of dynamical systems, chaos, and self-organization.

Funding of research in biological physics is difficult to appraise since it blends into that for medical research. Qualitatively there are shortages of funding on the basic physics side of the interface, particularly where new directions in physics first appear and in those areas of biophysics not associated with human health. Although the National Science Foundation is supposed to meet these needs, its program appears to be spread so thinly over a range of cell biology, biochemistry, physiology, and biophysics that the intimate relationship between biophysics and the rest of physics tends to be overlooked. New research patterns derived from modern advances in theoretical and experimental physics are not readily accommodated by the established support system.

Several current *scientific opportunities* in biological physics offer great promise: they are the possibilities to (1) develop basic physical understanding of biological macromolecules to the stage at which structure and function are quantitatively related; (2) understand the organization and basic molecular mechanisms of the brain; (3) establish working connections between genetic biotechnology and biophysics; and (4) explore theoretical analyses of biological systems as physical problems encompassed by general dynamical systems with potential for chaos and self-organization.

It is essential that the sizable growth of support for genetic control of protein structure and function be linked with fast and effective methods for determining those structures, particularly the important active sites. Thus biophysics and biotechnology need to become partners.

Research on the brain at the biophysical level represents a special opportunity at an exquisitely poised state of the field. The national need for progress in this area is all too well expressed by the health-care costs associated with failure of the brain's function, which exceed even the costs of its education. The opportunity for biological physics at the supramolecular, cellular, and multicellular levels is extensive. Physiology and neuroscience pose problems in biological physics that appear ready for sizable progress.

Although biophysics research can, like most lively research fields, benefit from substantial increases of research funding, great benefit could also accrue from some *restructuring of funding*. For example, funding of basic structural and theoretical research on protein and polynucleotide structure and function seems to have been following parallel, rather institutionalized disciplinary pathways in the chemistry, biology, and biophysics communities, whereas more interdisciplinary viewpoints are now more appropriate. It is recommended that systems be explored within the funding agencies to accommodate interdisciplinary research that draws from deep within physics far from the usual methods of biophysics. For example, in the biophysics of brain and nerve, the principal funding pathways seem to support either a rather institutionalized neurophysiology or the now fashionable fields of computer science, robotics, and automation. The now fruitful conjunction of these disparate viewpoints in interdisciplinary biophysics research seems to be difficult to fund or publish. New funding in this sort of area might provide for more innovative and productive interaction.

The conjunction of molecular biophysics and genetic biotechnology is shaping up as one of the greatest areas of opportunity for fundamental biophysical processes. Here the problem is that the two fields are so disparate that progress in establishing contact between them is very slow. The potential is sufficient that specific postdoctoral, sabbatical, or research-level funding to facilitate cross-disciplinary learning and experience is appropriate and should be made.

Interdisciplinary research in biophysics would be facilitated and stimulated by formation of a few *university centers for biophysics*. Interdisciplinary laboratories have revolutionized research in the materials sciences and in some special areas of biology. The sophistication of physical methods now used in biology makes experimental biophysics a demanding and expensive discipline even before the profound and complex questions of contemporary biology are addressed. Funding of centers for biophysics can bring together the potential of the latest experimental methods of both biology and physics through their central facilities. Collaborations between the associated biologists and physicists for advanced biophysical research and for interdisciplinary education stimulate experts from both sides of the biology-physics interface to productive and creative interactions.

Among the relevant physical and biological methodologies that can be exploited in this way are synchrotron radiation facilities for dynamic diffraction and EXAFS; digital image-intensified laser light and fluorescence microscopy; scanning transmission and microprobe electron microscopy; NMR in its high-frequency, pulsed, and multidimensional elaborations; single-channel electrophysiological recording; cloning in higher cells and plant cells; multidimensional fluorescence-activated cell sorting; and large-scale computing. Since the basic physics of complex cooperative systems is reaching the degree of sophistication at which it can begin to address biological systems as fundamental problems in physics, pure physics becomes an important component of such centers. Combining these capabilities in several interacting biophysical research communities is recommended to exploit the conspicuous opportunities in biophysics in our time.

In summary, this chapter describes a productive and lively interdisciplinary field that calls for permissive support patterns to promote innovation and diversity through multidisciplinary approaches to biological problems. Support of increased biophysical research in non-health-related biology as well as the biophysics associated with medicine would address perceived opportunities.

3

New Aspects of the Physics-Chemistry Interface

INTRODUCTION

The interface between physics and chemistry has been crossed so often in both directions that its exact location is obscure: its passage is signaled more by gradual changes in language and approach than by any sharp demarcation in content. It has been a source of continual advances in concept and application all across the science of molecules and atoms, surfaces and interfaces, and fluids and solids. Yet, in spite of this, the degree of direct, collaborative interaction between physicists and chemists in the United States, especially at universities, has remained surprisingly limited. These relationships have recently begun to grow, especially in the region of interdisciplinary overlap. This chapter outlines some of the science responsible for the changes, suggests the kinds of opportunities that these changes represent, and examines means of exploiting them for new science and technology.

In preparing this summary, we have borne in mind that the robust subfield of chemical physics has already been treated in detail in the report of the National Research Council Committee to Survey Opportunities in the Chemical Sciences (*Opportunities in Chemistry*, National Academy Press, Washington, D.C., 1985). We do not attempt to reproduce that discussion here, but rather we concentrate on the particular emerging interactions between chemists and physicists.

Much of the traditional interaction between physics and chemistry

53

has not been of the direct, collaborative nature. Rather it has occurred by the transfer of techniques or experimental and theoretical discoveries following their introduction in one field. In classical thermodynamics; in the quantum mechanics of atoms and molecules; and in x-ray crystallography, optical spectroscopy, and magnetic resonance, physicists have often laid the basic foundations; but both physicists and chemists have built on and enriched them to create new subfields. In statistical mechanics, irreversible thermodynamics, and some aspects of fluid mechanics, chemists have often made the fundamental discoveries that were taken up by physicists. These patterns of scientific transfer have recurred continually over the past century and do persist today. As important as they have been, however, they do not constitute true interdisciplinary science.

The advent of new instrumentation of unprecedented power has required chemists either to develop instrumentation in the traditional fashion of physicists or else to enter into close collaboration with physicists in order to address the microscopic properties of complex molecules, materials, and interfaces. For their part, condensed-matter physicists, propelled by instrumentation, theory, and technology toward materials of increasing complexity, find themselves often confronting questions requiring increased chemical insight and techniques for their answers. Finally, new complex materials displaying remarkable new chemical and physical properties have nucleated a coalition among synthetic chemists, physical experimentalists, and theoretical physicists from deep within their respective fields, far from the traditional interface. Each of these currents has already created new fields and produced new discoveries in established fields. As the principals continue to collaborate and to learn one another's language, there is reason to expect much more.

We concentrate in this chapter on a few examples chosen to illustrate some of these themes. Our emphasis on these particular examples does not necessarily reflect the actual proportion of research activity devoted to them but rather the importance of the new modes of interaction that they represent. The next section reviews the impact of advances in instrumentation (lasers, surface-science probes, neutrons, and synchrotron radiation) whose influence at the interface is rapidly evolving and is fostering stronger collaborations between chemists and physicists. The section that follows it treats the physics of the traditionally chemical subject of polymers and complex fluids, and the next section discusses the physics-chemistry synergism that has begun to generate organic electronic materials. The final section summarizes some of the nontechnical issues associated with research at the interface and presents some recommendations for dealing with them.

INSTRUMENTATION-DRIVEN COLLABORATION

Laser Science

One traditional area of interaction between chemistry and physics has been the realm of spectroscopy. Some of the most exciting advances occurred in these areas in the past 10 to 15 years, particularly as a result of major advances in instrumentation and measurement techniques but also as a result of advances in theory. The advent of the laser as a source of coherent, tunable, monochromatic radiation in the ultraviolet to the far-infrared regions of the spectrum has led to a number of important advances.

First, in gas-phase molecular spectroscopy it has become possible to test our understanding of the atomic and molecular hyperpotential to unprecedented levels, including aspects of intramolecular dynamics. The simple elegance of the coherent spectroscopies belies the immense scientific and technological challenges that had to be met to make them possible. Some of these experiments provide the scientific basis of such diverse technologically important areas as laser isotope separation, detection of pollutants, control of chemical processing, and the important science of combustion and the intricate details of the mechanism of surface chemical reactions.

Further advances in laser spectroscopy concentrate on probing liquids and polymers as well as molecules bound to surfaces at complex interfaces. Our understanding of surface-enhanced Raman spectroscopy has made it possible, under the right circumstances, to probe surface-bound molecules specifically at the liquid-solid interface.

The important development of supercooled molecular-beam spectroscopy not only simplifies the study of molecular spectra but makes it possible to explore problems in solvation and material synthesis. It now provides a new focus for all aspects of molecular spectroscopy, from laser-isotope separation to the production of naked metal clusters that simulate molecular surfaces. The next 10 years will see much new science generated here.

The availability of intense, monochromatic coherent light that can excite specific vibrational modes in specific complex molecules has spurred the rapid evolution of laser chemistry over the past decade. New developments include multiple photon dissociation, which is important not only for isotope and isomer separation but also to probe fundamental aspects of unimolecular and bimolecular reaction rate theory. Laser chemistry has also generated significant insight into the flow of energy in molecular systems. The availability of efficient ultraviolet lasers has stimulated work in radical chemistry leading to new and more efficient photochemical processes to produce circuits and materials.

A new frontier in laser chemistry involves shorter times of excitation and further exploration of the problem of energy equipartitioning in chemical reactions. During the past 15 years the shortest optical pulses that could be generated have decreased from picoseconds to about 8 femtoseconds long (about four vibrations of a visible light wave). With pulses this short the dynamics of almost any chemical or electronic process can be directly probed. Laser chemistry, spectroscopy, and theory are also having a profound effect on our understanding of reaction dynamics and on our ability to detect transition state species. This should lead to the capacity to monitor and control processes on a practical basis.

The spectroscopy of solids has been important not only for examining the electronic and vibrational structure of solids but also for studying the subtle effects of solid-state phase transitions. As we note in the next section, these spectroscopies have been extended to study surfaces and the interactions of molecules with surfaces.

Laser physics will play an extremely important role in all these areas, especially as the availability of laser sources extends well into the ultraviolet and into the soft-x-ray regime. Efficient new devices will open up new areas in the chemistry and physics of molecular ions in the gas and liquid phases as well as at interfaces. Although there is considerable speculation about the role of ions in chemical reactions, such as combustion, soot formation, nucleation, electrochemical reactions, and even catalytic processes, there has been little opportunity to explore ions experimentally in anything but gaseous systems. Significant theoretical advances are likely in these areas as well. Much early work using multiphoton ionization has stimulated the field, but its advance is limited by lack of readily available sources.

Surface and Interface Probes

During the past decade we have witnessed spectacular strides toward achieving a molecular-level understanding of chemistry at interfaces. Ultrahigh-vacuum and surface-preparation techniques permit exploration of the chemistry on nearly perfect single-crystal faces that remain free of contamination for the duration of the experiment. Powerful analytical techniques characterize structural and electronic properties of the surface and chemically active adsorbates. Ultraviolet photoemission spectroscopy, analytical electron microscopy, low-energy analytical diffraction, and electron energy-loss spectroscopy are just a few of the new tools that have emerged during this period. With the advent of synchrotron sources, new surface-sensitive techniques such as inverse photoemission spectroscopy and photoelectron diffraction have permitted unprecedented measurements of the electronic states of surfaces and adsorbates. Molecular-beam techniques have dramatically enhanced our ability to control and follow fast chemical changes on the surface. Lasers are utilized both for diagnostic purposes and to initiate or alter chemistry on the surface. Hand in hand with these experimental advances, the theory of electronic structure and dynamics has developed rapidly and, in some cases, provided valuable leadership to experiment.

In molecular-beam scattering experiments, temporally modulated monoenergetic beams of reactant molecules impinge upon a well-characterized surface; and the angular, time-of-flight, and internal energy distributions of gaseous products are detected. The ability to observe nascent reaction products free from subsequent gas-phase or surface encounters on a microsecond time scale, with control over incident gas conditions, coverages, and surface temperatures, has proved useful in uncovering the mechanisms and rates of elementary reaction steps. Furthermore, the use of lasers to prepare reactant molecules in specific initial quantum states, and to detect the final quantum states of products, has for the first time permitted exploration of the flow of energy during elementary surface reactions. These studies are beginning to reveal such energy mechanisms as the nature of reaction barriers, the dynamics of surmounting the barrier, the time scale for energy equilibration, and the nature of sequential reaction and desorption steps. This knowledge, in turn, may lead to optimal design of surface composition and topography to produce a desired chemistry.

Exciting progress notwithstanding, there are serious limitations to our current experimental capabilities. Modulated molecular-beam studies with laser diagnostics characterize gas-phase products but do not yet adequately concurrently monitor adsorbed species. The general requirement is for a time-resolved spectroscopy that can follow the rapid changes on the surface as they evolve during a modulated molecular-beam cycle. A high-resolution vibrational spectroscopy would be of the most value because of its potential to reveal the identity, bonding, and local environment of transient surface species.

Recent advances in time-resolved electron-energy-loss spectroscopy promise to fill this need and to make possible a new depth of inquiry in the near future.

The role of defect, or active, sites in surface chemistry has been discussed for decades. Recent molecular-beam studies have documented dramatically the pervasive influence of defects on low-coverage kinetics; and static spectroscopies such as angle-resolved electron-stimulated desorption, extended x-ray absorption fine structure (EXAFS), and nuclear magnetic resonance have characterized the active sites in a number of catalytic materials. Time-resolved vibrational spectroscopy now offers the possibility of following, through changes in spectral features, the diffusion of adsorbates to active sites, the formation of intermediates and products, and subsequent escape from the sites. Another exciting new tool, the scanning tunnel microscope, will have an extraordinary impact in this area as well. It may soon be possible to image individual atoms of an adsorbate-defect unit to determine definitively the identity and bonding of active sites.

The precise data now attainable on rates, mechanisms, and energetic constraints of elementary reaction steps provide unprecedented information about the potential energy hypersurfaces that govern the motions of rearranging atoms. Significant progress has been made by theoretical chemists and physicists in semiempirical and first-principles calculation of such potential energy hypersurfaces and in the tracing out of classical or quantal trajectories of atomic motion over the hypersurfaces through the course of a reactive event. Theory has provided critical insights into geometric, electronic, and vibrational properties of adsorbates, into the promotion and inhibition of reactivity on metals by electron-donating or -withdrawing elements into the rates and energy distributions of thermally and laser-desorbed molecules, and into the involvement of phonons and electron-hole pairs in energy transfer. It is encouraging to note that the relationship between band-structure physicists and small-molecule quantum chemists has evolved rapidly from the neglect or hostility of 10 years ago to a current state of mutual appreciation. We look forward to future cooperation that fully exploits the strengths of both approaches.

Another emerging field is the physics and chemistry of small clusters. It is now possible to prepare clusters of from two atoms to several hundred atoms with a precise number and a narrow size distribution. Methods for their structure determination are rapidly being developed. Organometallic clusters in gas, liquid, or crystal phases; supported catalytic particles; suspensions of small colloidal semiconductor particles; and naked gas-phase semiconductor and metal clusters are all being actively studied. Data on metal-metal and ligand-metal bonding obtained from studies of organometallic clusters will provide an invaluable foundation for understanding bonding to metal surfaces. The chemical and electronic properties of naked gas-phase clusters, synthesized in free-jet expansions, constitute an ideal testing ground for electronic structure theories as well as providing new insights into size effects and the approach to bulk behavior. Although currently the major focus in cluster research is synthesis and structural characterization, the evolution toward an emphasis on the chemical properties of clusters that is now beginning promises substantial achievements in understanding, in addition to the possibility of accomplishing highly specific chemistry on clusters of selected size and structure.

The next decade promises to clarify substantially our previously murky picture of chemistry occurring at the gas-solid and liquid-solid interfaces. One reason for great excitement here is the potential impact on such processes as corrosion, oxidation, combustion, passivation, adhesion, heterogeneous catalysis, crystal growth, molecular-beam epitaxy, chemical vapor deposition, and plasma etching. Ultimate optimal control of these processes will require a knowledge base that extends to the molecular level. Single crystals under ultrahigh vacuum are not, of course, found in commercial plasma etching chambers or catalytic reactors. But several recent studies have demonstrated a remarkably close correspondence between chemical reactivity on single crystals and on surfaces of practical interest. Experimental techniques are being developed to probe high-pressure surface and liquid-surface chemistry. Laser microchemical techniques already have important applications in fabrication of electronic devices. Major progress is forecasted not only in our understanding of interfacial chemistry but also ultimately in our ability to tailor interfaces specifically to achieve a desired chemical behavior.

Neutrons and Synchrotron Radiation

Two especially powerful probes of the microscopic properties of complex materials have their origins at the frontiers of physics. Synchrotron radiation was originally used to investigate elementary-particle interactions, and use of controlled neutrons arose, of course, from the investigations of nuclear physics. Both, however, are now often used for the investigation of the microscopic properties of complex materials. Synchrotrons provide over one million times more intensity of x rays and ultraviolet radiation than is available from laboratory-based x-ray tubes or discharge lamps. The high intensity and collimation of synchrotron radiation make it ideal for studies of small amounts of material, such as surface adsorbates or novel compounds available only as very small crystals. Neutrons have special sensitivities to lattice vibrations, magnetic excitations, and hydrogen. This evolution has been directed primarily by physicists involved in the emergence of condensed-matter physics over the past four decades, but from the beginning it has also had a significant chemistry component. The combination of increasing materials complexity and instrumental sophistication often requires partnership between physicists and chemists with complementary skills. Indeed, today interdisciplinary teams, sometimes from different home institutions, working together at large off-site facilities is the norm rather than the exception. The advances in characterization capabilities provided by these large facilities and the richness and control of novel materials are already producing significant advances and promise many future scientific and technical innovations.

Among the significant accomplishments of neutron scattering are the study of tunneling modes in chemical systems such as deuterated solid methane, the determination of single-polymer-chain configurations in bulk and composite polymers, and the chemical crystallography of several solid-state and catalytic materials.

The future opportunities for chemical research using neutrons involve both dynamical and structural studies. Low-energy cold neutron spectroscopy can now be used to study rotational processes, tunneling phenomena, and ionic and molecular diffusion mechanisms in both homogeneous and heterogeneous chemical media. These studies should provide valuable insight into the behavior and activity of catalysts, chemical adsorbates, and intercalated and layered materials. High-energy neutron studies will test dynamic models of molecules bound to chemical systems such as adsorbates on supported catalysts as well as the dynamical behavior of molecular liquids and colloid and micellar complex solutions.

Structural studies, particularly powder diffraction techniques to study materials that cannot be prepared as single crystals, will be extended by both source and detector developments to the study of more complex and metastable systems under various temperature and pressure conditions. These include studies of ionic conductors, conducting organic solids, ceramics, and catalysts.

The unprecedented brightness of synchrotron radiation, its tunability throughout the electromagnetic spectrum, and major advances in techniques and instrumentation have resulted in a revolution in analytical capability for complex materials. The three major techniques are

EXAFS, photoemission, and diffraction. EXAFS has been used to study near-neighbor atomic distributions in complex enzymes, amorphous solids, atomic absorbates, and thin epitaxial layers on crystalline solids. This technique often provides complementary structural information to that provided by other techniques such as x-ray diffraction, but it is sometimes a unique tool, especially for studies of the coordination about an element at low concentration. For example, catalytic systems under real reaction conditions have been studied with tens of parts per million sensitivities for the catalytic site. Photoemission spectroscopy has provided detailed understanding of the electronic structure of many interesting systems, such as gaseous molecules, adsorbates, reactive surfaces, and bulk solids. The high brightness of synchrotron radiation can be used to determine by standard crystallographic techniques the structure of important complex materials such as zeolites, which can be grown only in 1-10-μm sized crystals. Also, the techniques of surface diffraction and surface EXAFS are providing for the first time reliable structural information on reactive surfaces. Submonolayer sensitivities have been demonstrated for adsorbate systems.

The future promises even greater brightness of synchrotron radiation with the use of special magnetic insertion devices called undulators. This will further extend all the capabilities described above. In addition, we expect the development of an x-ray microscope and other imaging techniques that will permit study of heterogeneous chemical systems in situ. Real-time experiments in which the structural changes associated with reactions are followed will become a possibility. The use of storage ring technology for the free-electron laser opens the possibility of developing new tunable laser sources into the ultraviolet that will provide a formidable new probe to the chemical physicist.

The collaboration between physicists and chemists that is occurring at major synchrotron facilities will result in many future advances in both fields. The major neutron and synchrotron facilities are being exploited by a new team of forefront physicists and chemists. The two cultures are merging, providing an integrated study of many complex materials of interest to both.

POLYMERS AND COMPLEX FLUIDS

Over the past decade condensed-matter physicists have become increasingly concerned with problems involving macromolecular systems. In this area the traditional boundaries between chemistry, physics, and, to some extent, biology have become blurred. The development of this convergence has, however, sometimes been hindered by the insular nature of typical university departmental organization, providing part of the rationale for a recommendation of our final section.

Recent experimental and theoretical developments have stimulated a renaissance of interest and activity for both random and organized macromolecular structures. On the experimental side, the emergence of advanced structural probes such as neutron and x-ray scattering enable us to investigate conformations on a broad range of length scales ranging from atomic to optical dimensions. Selective labeling and contrast matching permit molecular specificity. For example, selective perdeuteration of a small fraction of polymer chains in an entangled network yields single-chain conformations. The brightness provided by synchrotrons helps us to study weakly scattering structures, such as surfactant monolayers, liquid-crystal films, and surface-adsorbed polymers.

On the theoretical side, application of modern critical phenomena and liquid-state theories to polymer solutions, colloidal suspensions, and microemulsions has begun to generate a new level of understanding of these complex systems. The discovery that the excluded volume interaction in polymer solutions can be mapped onto an $n = 0$ field theory and the discovery of related scaling behavior have yielded a reasonable understanding of entangled transient networks. The van der Waals theory for binary mixtures has been usefully applied to model the phase behavior of microemulsions. Similarly, percolation concepts have been extended to produce generic gelation phase diagrams. These are but a few examples of the penetration of critical phenomena theory into the field of macromolecular structure. Liquid-state theory is providing a detailed interpretation of the correlations in colloidal suspensions and, augmented with hydrodynamic interactions, of both linear and turbulent response.

The field of liquid crystals epitomizes the interplay among sophisticated diffraction studies, creative synthesis, and phase transition theory that has generated a deep understanding and appreciation for the role of symmetry in these exotic fluids. These advances have provided spin-offs to other areas, such as two-dimensional melting and even lattice gauge theories in elementary-particle physics. Multicomponent systems often exhibit some degree of self-assembly, which naturally gives use to materials with unusual interfacial properties. The basis required to elucidate macromolecular interfaces and the related special surface states such as prewetting and critical wetting is now rapidly emerging.

Many macromolecular systems exhibit short-range order and long-range disorder; examples are lyotropic phases, micellar and colloidal crystals, birefringent microemulsions, and blue phases. Over the next few years, a unified understanding of this weak ordering will develop, with important implications for the consideration of biological self-assembly. Research opportunities abound in the elucidation of aqueous solutions for general polyelectrolyte problems, membrane structures and activity, and emulsification. The relatively unexplored area of conjugated chain solutions will receive much attention.

As understanding of complex physical-chemical systems develops, some measure of control will follow. The relationships between molecular structure and macroscopic properties should enable us to develop algorithms for constructive interplay between chemical synthesis and physical properties for predictive chemical and biological engineering. Powerful computational methods and tools are already making important contributions to the properties of fractal and more local random structures.

If the arbitrary chemistry-physics-biology barriers can be effectively broken down and continued growth in our technological capabilities is maintained, the macromolecular interface should experience significant development in the next decade.

ORGANIC ELECTRONIC MATERIALS

Before 1972, no known organic compound was a metallic conductor of electricity. Now there are two broad classes of such conductors whose electronic and physical behavior can be quite different from those of conventional materials: the charge-transfer salts [e.g., tetrathiafulvalene (TTF)] and polymeric systems such as doped polyacetylene. Among the most prominent examples are organic superconductors, nearly one-dimensional metals with unique collective ground states, polymers whose conductivities can be controlled over 12 orders of magnitude, and nonlinear optical materials of exceptionally high efficiency. With the new materials have emerged new physics and new chemistry. Equally significant are the synergistic interactions that they have engendered among synthetic organic chemists, experimental physicists, and theorists.

Conducting Molecular Crystals

The charge-transfer salts are molecular crystals in which stacks of planar aromatic molecules have been partially ionized to form conducting paths along the stacks. Because of their anisotropic structures, the charge-transfer salts display all the instabilities long predicted for hypothetical one-dimensional metals—charge-density waves, spin-density waves, complete localization in the presence of disorder, and large fluctuation effects. Because they are narrow-band materials (typical conduction bandwidths are less than 0.5 eV), they also display phenomena associated with the near breakdown of the metallic state—polaron formation, magnetism, strong Coulomb interactions, anomalous optical properties. Finally, many of the properties are quite sensitive to chemical changes. Through relatively minor manipulations of molecular architecture such as substitution of selenium for sulfur, or of fluorine or a methyl group for hydrogen, we can pass systematically from magnetic insulators through semiconductors to semimetals and metals. The culmination of this process has been the development of the first organic superconductors. In the future, it may be possible to design organic molecules to produce specific solid-state properties directly.

Conjugated Polymers

The second, somewhat newer, class of organic conductors consists of doped conjugated polymers such as polyacetylene and a variety of polyaromatics. These can be prepared as large-area films of reasonable molecular weight. They are usually insulators in their pristine state, but their conductivity can be varied over as many as 12 orders of magnitude by controlled doping. At low doping levels some fascinating physics has been attributed to neutral radical defects or charge carriers combined with associated lattice distortions to form mobile solitons, polarons, and bipolarons; the theory of such states is in fact similar to some of those developed by field theorists to explain the nature of fundamental particles. At higher doping levels a semiconductor-to-metal transition occurs.

The nonlinear optical responses of extended conjugated systems, especially in polydiacetylenes, more specifically the third-order nonlinear optical susceptibilities, are 3 orders of magnitude greater than those of conventional materials. Spectroscopic studies of the excitations in polyacetylene have shown that photoexcited carriers are produced by the trapping of a few percent of the electrons or holes at neutral soliton defects, while the rest recombine from an exitoniclike state. This behavior is quite novel and different from the behavior of such excitations in inorganic semiconductor systems. A full understanding of the excitation spectra, carrier dynamics, and optical nonlinearities in polymeric systems will likely emerge in the near future.

From a materials point of view, the polymers are probably more attractive than the charge-transfer salts by virtue of their greater strength, but at this point only limited processibility has been achieved. The development of fully processible forms remains a central synthetic problem but may open the way to applications such as lightweight conductors to antistatic and radar-absorbing materials.

More broadly, it is important to recognize that the class of conducting polymers is still small and simple compared with the range of synthetic possibilities. As better-characterized materials are prepared and fundamental understanding grows, the power of organic synthesis ought to be directed to produce specific electronic properties in these materials and perhaps some technologically important applications.

Molecular Assemblies

As biological examples demonstrate dramatically, organic matter is capable of structural and functional self-organization on a remarkably complex level. The organic materials in which interesting electronic properties have been attained—molecular crystals and simple conjugated polymers—exploit this capability only in its most primitive forms. As a next step, it is natural to consider the prospects for incorporating such properties into more complex organic assemblies.

The simplest set of examples consists of noncrystalline charge-transfer complexes, such as dispersions of tetracyanoquinodimethane (TCNQ) in processible polymers or polycrystalline films of organic semiconductors. The former exhibit unexpectedly high electrical conductivities without compromising the mechanical properties of the polymers, and the latter exhibit reversible subnanosecond switching phenomena in response to light or electric fields.

Still under development are electrically conducting liquid crystals and polymer gels. These are of interest not only for the expectedly new physics of interaction between electronic and mechanical or flow properties but also for the prospect of controlling electronic behavior through various applied stresses or orienting fields.

Especially promising are techniques for preparing monolayer molecular films and complex multilayer composites. This is usually accomplished by Langmuir-Blodgett techniques, but other approaches such as solution epitaxy have also proved fruitful. In some cases monomer films can be polymerized in situ for strength or conductivity. Langmuir-Blodgett films may have interesting active electronic or optical properties of their own. Such assemblies have been used to elucidate mechanisms of light-induced charge and energy transfer by systematically varying the separation between active molecules and as light-trapping nonlinear optical devices. Little has been done, however, to introduce electrical conductivity into the films and to construct sandwich or multilayer superlattice structures with new composite electronic properties. This seems an area ripe for further development.

The organic electronic materials of the past decade have stimulated our curiosity. They have introduced high electrical conductivity and associated electronic properties into new parts of the periodic table, and they have exhibited new phenomena that have helped to generate new intellectual thrusts in condensed-matter physics. They have drawn organic chemistry, physical experiment, and condensed-matter theory together in new ways. Yet, for all that, they have barely begun to address all possibilities for the organic synthesis or the special material properties of organic structures. With appropriate support, we expect their second decade to be highly productive.

RECOMMENDATIONS

Education

It is our sense that neither in chemistry nor in physics departments have formal education patterns adequately accommodated areas of overlap, nor have they kept pace with the new developments at the interface. Barriers remain in language, orientation, and traditional lack of collaboration between scientists in the two fields.

We recommend that transdisciplinary courses be designed to teach physics and materials science, particularly condensed-matter physics, with full rigor to chemistry students, in ways that make connections with what the students have already learned through their grounding in chemistry.

In physics and the material sciences, there is need also for the language of chemistry but more importantly for learning of chemical information and understanding of chemical reasoning as well as exposure to the complexity of real materials that makes up much of materials science.

We recommend the design of courses that take advantage of the basic background of physics students to teach them chemical principles and reasoning at their full elegance for complex materials and their variation across the periodic table. In addition, exposure of physics students to the properties of real complex materials and the ways in which they are likely to differ from idealized models is essential.

Both in chemistry and in physics, such courses are useful on the *graduate level*, but their content will be of maximum benefit when it is *integrated into the undergraduate curriculum from an early stage.*

Academic Research

The usual organization of academic physics and chemistry departments does not encourage the kind of interdisciplinary research that we have described. Although chemical physics is a well-established component of most chemistry departments, synergistic interactions between physicists and chemists far from the traditional interface have usually occurred in spite of departmental structures rather than because of them. What is needed is not necessarily a major realignment of departments: it is important that collaborators from both sides of the interface continue to draw intellectual sustenance from their parent disciplines. Rather, the trend toward extradepartmental institutes as centers of interdisciplinary activity should be encouraged, both organizationally and financially. It is particularly important that university hiring patterns attach priority to this kind of enterprise.

Further, at least in areas of overlap, faculty, staff, and postdoctoral positions in a given department should be filled by candidates from both disciplines or by candidates with joint appointments. This method is being actively pursued in Europe—especially in France—as well as in Japan to enhance direct interdisciplinary science. Indeed, major industrial research and development laboratories often are organized in interdisciplinary teams. Many of the advances in those laboratories could not have occurred if they were organized by discipline, as they are in universities. Students already familiar with such an interdisciplinary approach would presumably be better prepared for the industrial environment.

Funding

Funding agencies have been both enthusiastic and sophisticated in their support of research at the physics-chemistry interface. On the other hand, it is important to recognize that interdisciplinary research is apt to be collaborative and that in such cases the levels and patterns of traditional single-investigator funding may no longer be appropriate. This is especially true of the newer kinds of synergistic programs involving synthesis, physical characterization, and theory.

Funding agencies should give special attention to mechanisms for maintaining the minimum funding levels necessary to sustain interdisciplinary research involving multiple principal investigators, often in different departments and institutions.

This entails the support not only of personnel and supplies but also of small instrumentation and computing facilities.

SUMMARY

While the importance and future contributions of physics and chemistry as separate disciplines and their traditional interfaces will certainly continue, there are new opportunities for scientific and technological interactions that will grow relatively in importance. Interdisciplinary, multi-investigator investigations of complex materials by physicists and chemists have grown rapidly over the past 5 years, with many accomplishments to their credit, some of which have just been described. The next 10 years should see not only a continuation of this growth but a blossoming into a well-established pattern that has significantly contributed to our understanding of complex material systems with significant associated technological benefits. If the potential of these interactions and those of the physics-chemistry interface in general are to be realized fully, education, research, and funding patterns should be modified to continue to nourish this extremely important interface.

4

Physics and Materials Science

INTRODUCTION

Many of the most significant applications of physics have occurred at or near its interface with materials science. The changes in our society that have followed the invention of the transistor and the subsequent emergence of ever more sophisticated computers and scientific instruments seem as profound as those brought about by the Industrial Revolution. It is evident throughout this Physics Survey that these developments, in return, have had a major effect on the style and methodology of research in physics, most especially in materials science itself. New techniques and new points of view in this interdisciplinary area are posing interesting challenges both for scientists and for policymakers.

In this chapter we focus attention on some of the most basic scientific issues that arise at this interface. That is, we are concerned primarily with the way in which materials scientists acquire the new fundamental insights upon which new technologies are based. Applications of the current state of the art in this area are described in many of the chapters that make up this volume. There is, of course, no clear separation between pure and applied research in materials science. New fundamental understanding can have immediately foreseeable practical consequences, and, conversely, problems arising in detailed technological applications can often be seen to have broad basic

69

significance. This intertwining of basic and applied research will be a central theme in what follows. It may be useful at the outset to distinguish two complementary trends in modern materials research. First and most visibly, a great deal of attention is being paid to new materials, to new methods of preparation, and to new modes of analyzing materials as based on novel understanding or novel interpretations. Second, perhaps less visible but equally important, is the growing realization that many of the long-standing problems of materials engineering are, in fact, scientific questions of deep fundamental significance. The problem of metallurgical microstructures, to be discussed in more detail below, is a prime example of this trend. Here, new interest in an essentially macroscopic problem of old-fashioned classical physics has been stimulated by its apparent relation to problems of nonequilibrium pattern formation in a wide range of physical and even biological situations. Of course, both of these trends owe their vitality to modern instrumentation and computational capabilities. The new convergence of interest in interdisciplinary classical problems is specially conditioned on our growing ability to deal with the mathematics of complex nonlinear systems. In the long run, new fundamental discoveries emerging from the latter kind of research may turn out to be the most important of all. In the second section of this chapter, we present some of the outstanding highlights of the interactions that have occurred in the past at the physics/materials-science interface.

A common theme running throughout this volume is the involvement of materials in the impact of physics on other disciplines and areas of application. Many of the exciting interactions between physics and materials are covered in other chapters. In the third section of this chapter we develop these connections and point out where the reader should look in other chapters for details.

In the last two sections of this chapter, we describe two special areas of ongoing research that are not covered in other chapters. A section on amorphous and disordered systems is primarily an illustration of problems encountered in the study of novel materials, but it also has underlying elements pertaining to unsolved fundamental questions in statistical physics. Then we describe current efforts by physicists to understand the genesis of metallurgical microstructures. We conclude by offering some observations about institutional problems facing this field.

HISTORICAL HIGHLIGHTS

The interface between materials science and physics, especially condensed-matter physics, has been a fruitful area of scientific research throughout the modern history of both disciplines. The following selected highlights are of both historical and current interest. Note that each of these developments remains relevant to materials research in the 1980s.

The need for a systematic description of the various equilibrium phases of matter provided much of the motivation for the development of thermodynamics and statistical physics in the nineteenth and early twentieth centuries. Conversely, these theoretical advances provided the basis for exploration of the mechanical and thermodynamic properties of complex substances, and the resulting information continues to be essential for the design of new materials.

One of the classic problems in materials science involves the strength of materials, that is, the way materials deform and break when they are subjected to stress. Since the 1930s, quantum physics has provided the basis for calculations of the cohesive forces between atoms in solids. A major advance came with the understanding that mechanical behavior of real materials depends on the presence of defects, especially dislocations whose motions permit layers of atoms to slip over one another during large-scale deformations.

Defects in materials dominate many of their properties. For example, point defects such as vacancies and interstitials are responsible for solid-state diffusion. The defects or impurities in ionic crystals often determine the color of the crystals. Defects and dopants in semiconductors control their electrical properties. The dislocations that can be generated by deformation in semiconductors will interact with other defects and impurities to destroy devices. Many important questions remain unanswered. Although many of the defects in metals can be adequately described by Ping-Pong ball models, the properties of other materials are more complex. The bonding configuration in single defects in silicon, such as at a vacancy, at an interstitial, at a dislocation, or even in the 7 × 7 structure observed on a clean [111] silicon surface are the subject of intense interest and controversy. And our understanding of defects in other semiconductor materials is primitive by comparison. Powerful new experimental methods such as capacitance spectroscopy and atomic resolution microscopy are being brought to bear on these problems. Figure 4.1 shows a direct image of a defect in silicon. There is an important role in this area for physicists, both theorists who can devise tractable models and exper-

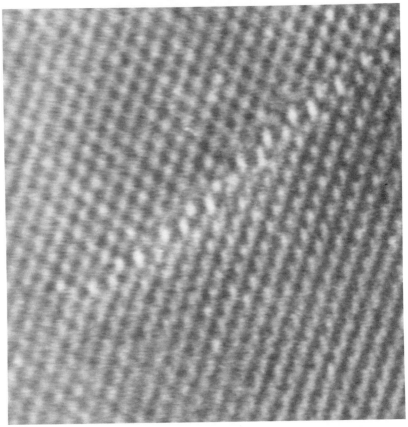

FIGURE 4.1 The electron microscope is capable of resolving, with atomic-scale resolution, the structure of defects in crystal lattices, such as this defect in a silicon crystal.

imentalists who may now be able to observe in detail what is actually happening.

The invention of the transistor in 1947 is the single most important example of the interaction between physics and materials science. Note that this extraordinarily practical invention was based directly on the quantum theory of electrons in crystalline solids—then not much more than a decade old. Note also that the realization of these quantum ideas was a materials problem; the transistor became a feasible concept when semiconducting materials of sufficient purity and crystallinity became accessible.

The discovery of a theory of superconductivity in the mid-1950s has led to deep new understanding of the properties of materials at low temperatures. Direct technological applications of superconductivity, such as in electric generators or transmission lines, are still speculative, but the implications of superconductivity for scientific instrumentation have been profound. Superconducting magnets that produce extremely stable, high fields have revolutionized the design of magnetic apparatus used in areas as diverse as medicine, geology, and elementary-particle physics. The same can be said of the family of extraordinarily sensitive measuring devices based on the Josephson effect. The design and fabrication of superconducting materials have generally been joint efforts of physicists and metallurgical engineers in which the physicists have contributed the fundamental insight that is needed in the search for new substances and the metallurgists have learned how to process these substances in ways that make them usable in real devices. An especially promising development has been the recent discovery of some superconducting organic polymers. Here the crucial collaboration has been between physicists and organic chemists. This potentially large and important area of materials research is just in its infancy as this report is being written.

The laser was viewed initially more as a scientific curiosity than as a technological innovation. Only the most visionary anticipated its widespread use in such diverse fields as surveying, manufacturing, medicine, and optical communications. Whereas the invention of the laser was based on a profound understanding of quantum physics, its development has depended on close cooperation between physicists and materials scientists, initially to provide uniformly doped single crystals for ruby laser rods and even now to develop novel semiconductor laser materials and structures. In turn, the laser has been used as a powerful probe of the optical properties of the new materials on which further important advances in quantum electronics depend.

A theoretical development of the 1970s that has profound implications for materials research is the new understanding of phase transitions that has been brought about by renormalization-group theory. This theory focuses primarily on critical phenomena, the large fluctuations that occur, for example, in a ferromagnet when it is heated to the point at which it loses its spontaneous magnetism or in a fluid when it is heated and compressed to the point at which the distinction between liquid and vapor phases disappears. It turns out that understanding what happens at critical points is crucial for the understanding of phase transitions in general. Armed with a first-principles theory of critical phenomena, physicists have begun to explore systematically the complex forms of phase equilibria that occur in multicomponent systems.

Results of work along these lines include the interpretation of what are now known as multicritical points in alloy phase diagrams and numerical techniques for predicting such diagrams by using only previously known atomic properties of the alloy constituents. A related development is the growing understanding of the kinetics of phase transformations, that is, the way in which transformation actually occurs during such processes as precipitation and solidification. This is an especially interesting topic about which we shall have more to say later.

Essential to any understanding of present opportunities for interdisciplinary research in materials science is an appreciation of the extraordinary progress that has been made in computation and scientific instrumentation during the past decade. The rapid growth in the speed and capacity of computers has enabled scientists to deal quantitatively with complex materials problems that seemed completely out of the range of possibilities only a few years ago. Of special importance to theorists is the growing availability of interactive computing systems with sophisticated graphics capabilities. The major significance of this research tool is only now beginning to be understood. The decade has also seen the emergence of physics-based instrumentation capable of probing properties of materials with a spatial resolution of atomic length scales (10^{-8} cm) and a temporal resolution of 10^{-15} s, faster than molecular vibration periods. Note, for example, the atomic-scale resolution of the electron micrograph shown in Figure 4.1. More detailed discussions of both computation and instrumentation may be found in Chapters 6 and 10.

THE PHYSICS/MATERIALS-SCIENCE INTERFACE

There is a continuum of activity between the physicist's desire to understand the physical world better and the materials scientist's desire to shape and control materials. As was illustrated in the previous section, many significant advances have depended on close cooperation between physicists and materials scientists, with physics pointing the way to new phenomena and materials science providing novel materials for realization. In many situations the boundary between physics and materials science is rather seamless and its exact location a matter of definition or personal preference. Advances in understanding of physical phenomena often result in new understanding of materials. These in turn can have important applications in diverse and unpredictable ways. So it is perhaps not surprising that many of the interfaces between physics and other disciplines, and many of the applications of physics, involve materials. This is a common theme running throughout this volume. In many of the chapters, there is a discussion of the materials that are involved in the various interactions and applications of physics. In this section we gather and regroup these interactions so that some of the dominant themes of materials science become apparent and so that the scope of this interaction can be seen. We do not discuss these topics in detail, but rather we present a matrix that the reader can use in referring to other chapters for detail. The natural groupings into which these interactions fall are new materials, new processes, chemical separation and analysis, surfaces, defects, instrumentation, and theory and modeling.

Each of these will be discussed briefly below. The interactions are not spread evenly across all chapters and groupings on this list, as can be seen from Table 4.1, in which are indicated the materials topics that are discussed in various chapters of this volume. It is an indication of the widespread nature of this interaction that nearly three quarters of the spaces are filled. It should be emphasized that this listing is not all-inclusive, just as the various chapters of this volume do not treat their topics exhaustively. The table and the rest of this section indicate the topics mentioned in the rest of this volume. Many readers will be able to provide additional examples.

In the next two sections of this chapter, we discuss in more detail two areas in which there are vigorous interactions between physics and materials science. In this section as well, we emphasize not only areas of past and present interaction but also some of the problem areas of materials science in which future cooperation can be expected not only to develop a deeper understanding of our physical world but also to improve our ability to shape and control materials.

New Materials

The contributions of physics to new materials and new uses of materials are discussed in several chapters of this volume. In microelectronics, modulation-doped structures, which were predicted to have dramatic effects on electron transport, have resulted in a new class of high-speed devices and in the observation of the quantized Hall effect. The concept of storing information in magnetic domains has led to the development of materials for magnetic bubble memories. New magnetic recording materials have been developed. Amorphous silicon and silicides are now used in silicon integrated-circuit fabrication. Amorphous silicon solar cells for the photovoltaic conversion of sunlight into electricity are discussed in Chapter 11, and insights into device operation continue to improve efficiencies. Optical communi-

TABLE 4.1 Interaction Between Materials Science and Physics Applications and Interfaces

Chapter Subject (Chapter No.)	New Materials	New Processes	Chemical Separation and Analysis	Surfaces	Defects	Instrumentation	Theory and Modeling
Applications							
Microelectronics (8)	x	x	x	x	x	x	x
Optical information technology (9)	x	x	x	x	x	x	x
Medical physics (13)	x			x			
National security (12)	x	x	x	x	x	x	
Energy and environment (11)	x	x	x	x	x	x	x
Instrumentation (10)	x		x	x	x	x	
Interfaces							
Biological physics (2)			x	x		x	x
Geophysics (5)						x	x
Chemical physics (3)	x	x	x	x	x	x	x
Materials science (4)	x	x	x	x	x	x	x
Computational physics (6)		x		x	x		x
Mathematical physics (7)							x

cations technology is based on the physical concept of the light pipe, but high-purity glasses and methods of fabricating them into fibers were needed before a practical system could be made. The other components of optical communications systems, such as semiconductor diodes, lasers, and detectors, are all based on semiconductor physics and are the subject of intense device-research activity at present. An important outgrowth of the development of fibers for optical communications has been the development of a wide variety of fiber-optic sensors. These are finding increasing applications in a variety of fields, including medicine.

Physicists have also contributed to the development of conducting molecular crystals, to conducting polymers, and to understanding of molecular assemblies, such as tetracyanoquinodimethane (TCNQ), as described in Chapter 3. Prosthetic materials represent an important area of interaction between materials science and medicine, which presents a major challenge in understanding the complex phenomena that occur at the interface between manufactured materials and living systems.

New Processes

Physics has contributed to processing technology primarily by providing new techniques, such as ultrahigh-vacuum technology and laser processing, and also by providing new insights into existing processes and new methods of analysis. Many of the thin-film techniques used in microelectronics have their origin in physics research. The most recent example of this is molecular-beam epitaxy, based on ultrahigh-vacuum technology and used, for example, for the fabrication of the modulation-doped structures mentioned above. As microelectronics continues to shrink its feature size and increase chip areas, physics-based methods such as ion implantation and laser annealing are of increasing importance in semiconductor-device fabrication. These processes have also been essential to the development of solar cells for energy conversion. The packaging of semiconductors presents a variety of important but widely ignored problems, for example, the role of surfaces in adhesion, to whose solutions physics is contributing. Many more opportunities exist in this area. New methods of solidification processing, such as rapid solidification and laser melting, are based on an improved understanding of crystallization processes and microstructure formation, as discussed below. Computer modeling of fluid flow, thermal diffusion, and mass transport are increasing the sophistication that can be brought to bear on the design of processing equipment.

Chemical Separation and Analysis

Physics has contributed heavily to chemical separation and chemical analysis through the development of novel methods and instrumentation. Lasers are now used extensively for spectroscopic studies of materials, not only for analysis but also for the exploration of the basic properties of materials including both the energetics and the kinetics. Time-resolved spectroscopy has been pushed to new limits with laser pulses as short as a few tenths of a femtosecond—comparable with atomic-vibration periods. The new regime of time opened up by this development is discussed in Chapter 9. Lasers are now being used to make ultrapure chemicals by photodissociation, and this technique is also used to prepare nuclear-reactor-fuel materials. The stringent control of purity necessary both in semiconductor materials and in glass for optical fibers is well known, and in both these cases there has been close interaction between physical analytical methods and chemical processing in the development of these ultrapure materials.

Surfaces

Surface science is an important branch of physics that has a direct interface with materials science. Surfaces are of greater importance than is often appreciated because most of the interactions between bulk materials and other substances take place at their surfaces. The properties of the bulk phases determine the thermodynamics, but the surfaces usually control the kinetics. The small volume of material at surfaces means that special techniques must be developed to study them. Surface science has contributed many such techniques, including electron-energy-loss spectroscopy (EELS), extended x-ray absorption fine structure (EXAFS) spectroscopy, ultraviolet photoelectron spectroscopy (UPS), low-energy electron diffraction (LEED), high-energy electron diffraction (HEED), and Auger electron spectroscopy, which have contributed to our understanding of surfaces. Two surface-dominated processes, catalysis and corrosion, are often cited to emphasize the importance of surfaces. Neither of these processes, catalysis (which generates billions of dollars in revenue annually in the chemical processing industry) nor corrosion (which annually destroys billions of dollars' worth of property), is well understood at present. For research into the dynamics of both catalysis and corrosion, advances are being made by cooperation among physicists, materials scientists, and chemists. The physicist brings the experimental tools of surface science and insights into surface electronic states; the materials scientist concentrates on the role of the structure, the grain size, and the surface and grain boundary segregation in these processes; and the chemist brings his knowledge of chemical reactions and a keen interest in applying the results. Surfaces and interfaces are also of great importance in microelectronics. For example, the properties of the semiconductor-oxide interface are of prime importance in devices, and the electrical properties of interfaces such as Shottky barriers are poorly understood but must be carefully controlled during device fabrication. Grain boundaries are internal surfaces that contribute importantly to the properties of metals, as discussed below in the section on Metallurgical Microstructures. Grain boundaries must be controlled in polycrystalline solar cells in order to achieve reasonably efficient solar-energy conversion. The study of surfaces continues to present a great challenge, both experimentally and theoretically, with great potential for practical applications.

Defects

Materials in single-crystal form are critical in many advanced applications. None of these crystals is perfect, and their properties of interest are often dominated by defects. The defects may be isolated point defects, or they may be line defects such as dislocations or even larger defects such as precipitates. The control of defects is often imperative in single-crystal applications. Physicists have contributed extensively to the understanding of the structure and properties of defects at the atomic level, of how the defects contribute to macroscopic properties, and also of the processes by which they originate. The defects in silicon have now been investigated in some detail, as discussed in Chapter 8, but even so, many problems remain. Defects in other semiconductors are even less well understood and under less control. For example, dislocation-free silicon crystals are grown routinely in production, but dislocation-free semi-insulating GaAs crystals (the most important semiconductor material after silicon) have not yet been grown, in spite of the considerable effort that has been expended. There is a significant challenge here to improve our basic understanding of defects, which will make a major contribution to the manufacture of microelectronic devices and solar cells. As is discussed in Chapter 11, the control of defects in solar cells has in the recent past contributed significantly to increased efficiencies.

Defects are important in non-single-crystal materials as well. For example, in Chapter 11 it is pointed out that radiation damage resulting from the generation of point defects by atomic displacements in reactor-wall materials and swelling resulting from the generation of helium from fission products in reactor fuel elements continue to be major problems.

Instrumentation

There is a separate chapter in this volume devoted to the contributions of physics to instrumentation, and it is clear that novel instrumentation for examining materials has been an important area of interaction between physics and materials science. This is exemplified by the ubiquitous electron microscope. This instrument is used extensively for examining biological materials, in microelectronics for the analysis of defects and devices, in energy conversion to examine the structure of solar cells, and in optical communications to study semiconductor lasers. X rays, which are a much older contribution of physics, are still widely used, not only in biophysics but as a major tool in many materials studies. Novel x-ray methods are also continually being developed, some based on new instruments such as the synchrotron and some based on novel analysis such as x-ray standing-wave studies of surfaces. Improved detectors have made energy-dispersive analysis of x rays a standard attachment for both scanning and transmission electron microscopes. EXAFS using synchrotron radiation is used for the study of surfaces of semiconductors and metals as well as biological materials. EELS is used for surface analysis and elemental analysis at high spatial resolution in the electron microscope. This method provides a unique analytical capability for using light elements in microelectronics as well as in biology, where tagged molecules can be followed. Laser-based instruments are used extensively for chemical analysis not only for process chemistry but also for analyses of the atmosphere.

Theory and Modeling

Theoretical and computational contributions to materials science are numerous and varied. Some are direct, but many are indirect. For example, Chapter 6 discusses new capabilities for modeling fluid flow. The results of such calculations are directly applicable to materials processing. But the general development of computational physics is a widely used new tool for the material scientist. The development of quantitative models is another major component of the interaction between physics and materials science. Many of the detailed models of material processes have been contributed by physicists, and, as is emphasized in the next section, the problems of materials science continue to offer fertile ground for this. An exciting area of current activity discussed in the next section is modeling of the pattern formation that occurs during the formation of metallurgical microstructures. But novel concepts in physics also frequently find application in materials science. For example, fractals are an important new theoretical concept with many potential applications in the areas of phase changes and microstructural development. Statistical-mechanics methods such as renormalization group theory used to model phase transitions are of great importance in chemical phase changes and in microstructure development. Crystal growth, which is the *sine qua non* of microelectronics, has been the subject of considerable modeling and computer simulation, so this complex process is now understood in considerable detail at the atomic level. Percolation concepts have been applied to phase separation and to the structure of glasses. The theory of phase transitions in liquid crystals has been related to the functioning of biological membranes. But there are many remaining challenges. For example, the structural and relaxation properties of glasses are poorly understood. The electronic properties of amorphous semiconductors are poorly understood. New theoretical insights are needed for understanding the structure and the electronic properties of defects in many materials. There have been many important theoretical contributions of physics to materials science, and the large number of remaining challenges makes this an exciting area of interaction.

AMORPHOUS AND DISORDERED MATERIALS

In the final two sections of this chapter we discuss in more detail two examples of the interactions between physics and materials science. These more extended discussions are intended to give some insight into how this interaction takes place. It will be evident that these areas of current interaction also present opportunities for future interaction. In this section we discuss amorphous and disordered systems, and in the next section, metallurgical microstructures.

One of the most interesting modern trends in materials science has been the attempt to understand the properties of amorphous and disordered materials. Until recently, most of the more visible achievements in the physics of materials have had to do with crystalline substances. In dealing with regularly ordered structures, one can talk with confidence about phonons, electrons, holes, Fermi surfaces, and other related physical terms that are now accepted as essential concepts by electrical engineers and metallurgists. However, many of the most interesting and useful materials are intrinsically irregular in their underlying structures. Some, such as glasses, are amorphous at the molecular level. Their constituent molecules are arranged irregularly, more like the molecules in a liquid than like those in a crystalline solid. Other materials, such as many naturally occurring rocks, are composites consisting of irregularly arranged small crystals. In fact, most real substances, including even metallic alloys, ordinarily are found in the form of multicrystalline composites. Thus, the problems of disorder are generally present whether one wishes to consider them or not.

Current technological interest in disordered materials is as varied as the materials themselves. Optically transparent glasses are being studied intensively for use as fast, accurate signal transmitters. Amorphous semiconductors are useful as photovoltaic materials and possibly for fast electronic devices. Metallic glasses have been found to have unique magnetic properties. They also have remarkable mechanical strength, as have certain artificially produced metallic composites. Plastics, of course, are amorphous solids made of organic polymers.

The relevance of fundamental physical problems to materials technology is nowhere more apparent than in this area of disordered systems—which is, of course, why we have chosen this area as a prime example of interdisciplinary interactions. One such problem, which has become familiar to physicists in recent years and which is described in more detail in the Condensed-Matter Physics volume of this survey, is the localization of electronic states. In crystalline solids, the electronic wave functions extend indefinitely throughout the system, and this quantum-mechanical uniformity is the basis for much of our understanding of the electronic properties of crystals—the freely moving electrons and holes referred to above, for example. In disordered systems, on the other hand, we now know that electrons of sufficiently low energy become quantum mechanically trapped by irregularities in their environment and that this localization has a profound effect on how those electrons behave in response to applied fields or other perturbations. Physicists have made important progress in understanding the localization effect and its implications, but the problem is by no means solved. We are not yet able to provide the materials engineers with a set of tools for dealing with electrons in amorphous systems comparable with those available for crystals. For example, electrons in solids repel one another by means of Coulomb forces.

These many-body interactions may play an even more important role in disordered systems than in ordered ones, but we do not yet have a sufficiently reliable and general theory of these effects. Another incompletely solved problem concerns the way in which localized electrons may be affected by thermal fluctuations of their environments in a disordered system. In a normal metal, these fluctuations scatter freely moving electrons and thus add to the electrical resistance. In a disordered material, on the other hand, such fluctuations may help electrons to escape from localization sites; thus, the resistance may decrease with increasing temperature. It would clearly be useful to have a quantitative understanding of such effects when one is developing amorphous materials for electronic devices.

An even more fundamental set of unanswered questions concerns the underlying geometry of amorphous structures. At present we do not have a systematic mathematical way of characterizing the degree of molecular order or disorder in such structures, and, as a result, we do not even know what rules to follow in carrying out thermodynamic analyses or applying the laws of statistical physics. The same structural order that we use to describe model crystals makes x-ray diffraction a powerful tool for determining their crystallographic structure. But the underlying disorder in amorphous materials makes this tool much less effective. The diffraction peaks from amorphous materials are quite broad and contain relatively little information about the local structure; or, rather, the local structural information is averaged over so many sites that detailed information is lost, and the resultant diffraction patterns cannot be used to distinguish among several distinct models of the local atomic structure.

Computer simulation has proved to be a valuable aid in understanding the local atomic arrangement of simple liquids such as argon where central forces are appropriate. Many properties have been described by computer simulations or in terms of models based on the simulations. Slightly more complex liquids—such as water—are much more difficult to simulate because of the shapes of their molecule, but simulations have given considerable insight into the local structure and into macroscopic properties. Computer simulations of glasses, on the other hand, quickly develop problems associated with long structural relaxation times. At some temperature, depending on the cooling rate, the structure of a glassy material is fixed. The glass transition temperature is the temperature below which a glass-forming material is essentially solid, and it is defined as the temperature at which the viscosity reaches 10^{13} poises on cooling. However, it is well known that most of the properties of a glass, such as its density, refractive index, and resistivity, depend on how fast the glass was cooled.

The structure and properties of a glass depend on the entire thermal history of the sample since it was last a fluid, that is, when it was last in a thermodynamic equilibrium state. This has led glass scientists to introduce the concept of the fictive temperature, which is the effective temperature at which the structure is fixed because of loss of mobility during cooling at a finite rate. The fictive temperature is defined as the temperature at which some property would have achieved its actual value had the sample been cooled infinitely slowly. This concept helps in rationalizing the measured properties of glasses, but in general the fictive temperatures for various properties, such as the specific heat, density, and viscosity, are not the same, since the various properties are not uniquely related to one another in a glass.

The most difficult questions about the amorphous state relate to structural relaxation. There are empirical descriptions of the structural relaxation of glasses that adequately describe the observed behavior, but, in general, the relaxation rates differ for various properties. The empirical models characterize the relaxation processes as occurring with a range of relaxation energies and rates, which presumably derive from the randomness of the structure. There is no adequate microscopic basis for these empirical descriptions, although they are all that the glass scientist has to work with at present. A particularly valuable method of gaining entry to the states available to glasses at low temperatures has been devised through studies of tunneling states at low temperatures in glasses. In a wide variety of glassy materials, coherent phonon echoes have been observed that can be used to measure dephasing times and relaxation processes in a regime that is not accessible to other measurement techniques. A useful description of the glassy state will have to be based on a systematic mathematical way of describing the structural disorder and will have to be couched in terms of kinetics, that is, of nonequilibrium concepts. Nonequilibrium statistical mechanics is one of the most challenging areas of theoretical physics.

The combined efforts of materials scientists and physicists working with a variety of experimental techniques, coupled with improvements in mathematical modeling and use of the increasing power of computer simulation, are expected to make significant advances in our understanding of this important and challenging class of materials. But this is a major undertaking, and it may well require novel experimental methods as well as the development of entirely new mathematical methods of describing the structure and modeling the relocation processes.

METALLURGICAL MICROSTRUCTURES

The typical appearance of the interior of a metallic alloy as seen, say, by observing a polished surface through an optical or electron microscope is that of a myriad of intricate patterns (Figure 4.2). The material is made up of crystalline grains whose typical sizes might be tenths of a millimeter. If the sample has not been deformed or annealed after solidification, and if the surface has been appropriately etched, one

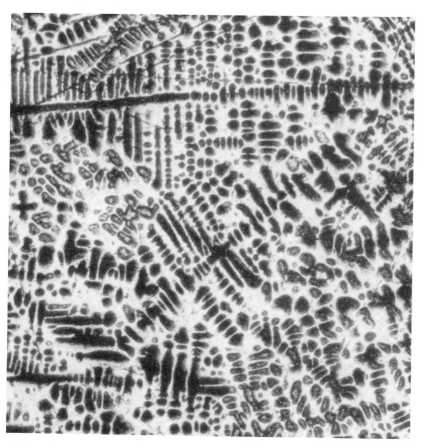

FIGURE 4.2   This complex composite structure in a metal alloy is an intimate mixture of two phases formed by dendritic growth of one of the phases during solidification.

may see dendritic—that is, snowflakelike—patterns on the scale of a few micrometers within each grain. These patterns are formed by the partial separation of the chemical constituents of the alloy that takes place during solidification. In alloys that have undergone heat treatments there will be evidence of further solid-state phase transformations. Precipitates may appear at grain boundaries or in regular patterns within the grains; altogether new crystalline particles may emerge; initially uniform solid solutions may give way to multiple new phases growing cooperatively in colonies of regularly spaced cellular or dendritic structures. The result may look like an array of feathers, a basket weaving, or a herringbone tweed.

These often complex patterns make up what is known as the microstructure of metallic alloys. (Similar patterns can be found in many other, nonmetallic, multicomponent materials.) Along with the structure of the grains themselves, it is this microstructure that determines the strength of the material, its electrical and magnetic properties, and how it behaves under heat treatment or mechanical deformation. For example, the speed at which dendrites grow into the molten alloy as it cools determines the sizes and shapes of the grain structure, because each dendrite determines the crystalline orientation of the material that solidifies around it. The composition variations created by the process of dendritic solidification or by later precipitation or growth of new phases control how easily dislocations move through the material and, thus, how easily the material can be deformed. Microstructural variations of composition or crystalline order provide nucleation sites for new thermodynamic phases that may appear during heating or aging and in that way control the properties of processed materials. For reasons such as these, the study of microstructures has always been central to metallurgical science.

A crucial feature of metallurgical microstructures is that they are intrinsically time-dependent, nonequilibrium phenomena. Like snowflakes, microstructural patterns are far from being stable, equilibrated forms of the crystalline materials of which they are made. If one waits awhile, snowflakes revert to compact ice crystals, and similar things happen in alloys. Complex patterns can occur only when the processes of formation take place in times much shorter than those required for the constituents of the system to find their most stable configurations. The latter times, however, may be years or even millenia. Whether one is dealing with fast or extremely slow processes, one finds qualitatively new and challenging problems in the physics of nonequilibrium phenomena. Important progress has been made in recent years, but many of the most fundamental of these problems remain unsolved.

One of the problem areas related to microstructures in which progress has been made during the past decade or so has to do with the atomic processes that occur at a solidification front, that is, at the moving interface between a growing crystalline solid and its melt. Under some circumstances, this interface is an atomically smooth crystalline facet. Such interfaces grow relatively slowly by means of mechanisms in which atoms attach themselves only at energetically favorable steplike irregularities. Under other circumstances, the interface is atomically rough and shows little if any evidence of crystalline anisotropy. We now understand that the transition between faceted and rough interfaces is, itself, a kind of thermodynamic phase transition. We are also beginning to understand how the atomic structure of an interface controls the growth rate and chemical composition of the emerging crystal. Theoretical techniques that have been playing major roles in these investigations have included renormalization group methods for studying the roughening transition and computer simulations of atomic kinetics at moving interfaces. Experimental methods have included light scattering using lasers to probe geometric features of nonequilibrium interfaces and a wide range of surface-scattering techniques using x rays, electrons, and ions, for example, to probe chemical composition.

A related area in which progress has been made, but whose story is still far from complete, is the study of pattern-forming instabilities of solidification fronts. Here we are talking about processes that occur on length scales much larger than atomic—tens of micrometers, for example—but for which the atomic nature of the solid-liquid interface is a controlling factor. A solid object growing into an undercooled or chemically supersaturated liquid characteristically undergoes a series of morphological instabilities in which small irregularities begin to grow increasingly rapidly, producing ever more complex patterns of deformations. It is this instability of simple shapes under nonequilibrium conditions that we now understand to be the generating mechanism for metallurgical microstructures as well as for many other kinds of patterns seen in nature (Figure 4.3).

Simple shapes of growing solids are unstable for reasons that have been understood for about 20 years. The mathematical theory of the onset of these instabilities is closely analogous to the theories of Rayleigh-Benard and Rayleigh-Taylor instabilities in hydrodynamics, an old problem in which there has recently been a major revival of interest among physicists investigating chaotic behavior. The metallurgical version of this problem, that is, solidification, seems much in need of the quantitative methods of investigation that physicists might bring

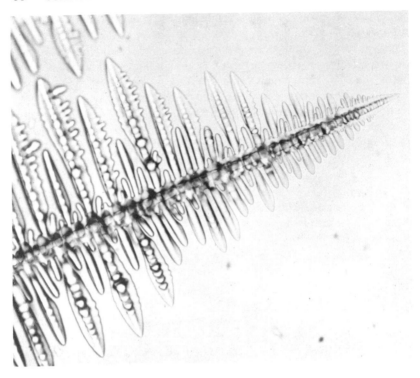

FIGURE 4.3   Pattern forming in nature leads to the intricate pattern of crystallization known as dendritic growth.

to bear on it. The need for new ideas is particularly acute as regards the obviously most interesting and important part of this problem, the question of how pattern selection occurs after the onset of morphological instability. Only during the past year or so have we begun to think that we know how to predict the growth speed of an isolated dendrite, the sharpness of its tip, and the spacing of its side branches. The new theory on which these predictions may be based, however, remains incomplete in some fundamental aspects. Generalization of these ideas, say, to arrays of dendritic structures or to cooperative growth of more than one solid phase as in directional solidification of eutectic solutions seems considerably more difficult—but even more important for practical metallurgical applications.

A principal reason that progress can be made in this area in ways that were not possible only a few years ago is the growing power of computers. The mathematical equations that describe even the simplest nontrivial models of pattern formation have novel features that do not yield easily to standard methods of analysis. We can now solve these equations numerically, and we can then use those solutions both to compare models with experiments and to gain fundamentally new insight into the mathematics. Another use of computers that is of special importance in this area is in the collection and interpretation of experimental data. An evolving solidification pattern presents the observer with a huge amount of information that must be stored and interpreted in ways that allow meaningful conclusions to be drawn from it. Digital image processing, for example, seems to be an extremely promising technique for this field.

A third large area of overlapping physical and metallurgical interest in microstructural problems pertains to phase transformations that occur after solidification. We have already alluded to the facts that solidified alloys generally are nonequilibrium structures and that the study of relaxation to thermodynamic equilibrium in such systems is at the very frontier of research in physics. A prime illustration of the state of the science in this area is the theory of nucleation and growth of precipitates. The chemical constituents of liquid or solid solutions often tend to separate when the system is cooled to sufficiently low temperatures. In dilute solutions, the minority constituents may come out of such a metastable solution in the form of small droplets, like fog. A relatively simple theory of the rate of nucleation of such droplets was developed more than 50 years ago; but it is only during the 1980s that the first direct tests of this theory are finally being carried out. Needless to say, there have been innumerable studies of nucleation in a vast range of fluid and solid systems during the past half century. But either these studies have been indirect, for example, measurements of cloud points rather than reaction rates, or else they have not been performed on carefully characterized systems in which the theory can be tested with confidence and without adjustable parameters. An especially useful technique in the latest investigations has been to look at model systems in which nucleation occurs near a critical point. Because of the progress that has been made recently in our understanding of equilibrium critical phenomena, it has become possible to use near-critical mixtures for accurate studies of nonequilibrium behavior. This use of physically well-characterized model systems for the study of fundamentally important materials processes is likely to be a recurring theme in the next decade or so.

As in the case of pattern formation, many of the most interesting problems having to do with the kinetics of phase transformations remain unsolved. A few examples should make it clear that there are excellent opportunities here for interdisciplinary research. Although some aspects of the nucleation problem seem to be coming under control, we do not yet have a reliable, quantitative method for predicting the actual numbers and sizes of droplets as functions of time after quench. In more technical language, we do not yet have a method for combining theories of nucleation with theories of growth. A related problem has to do with the behavior of systems that are quenched into thermodynamically unstable, as opposed to metastable, phases. This class of processes in phase-separating mixtures is usually called spinodal decomposition, and it is quite common in metallurgical alloys. Considerable progress has been made recently, again in conjunction with experiments on near-critical fluids, in understanding the early stages of spinodal decomposition; but as in nucleation, the problem of extending early-stage theories to quantitative predictions of the coarsening of later-stage precipitation has proven difficult to overcome. Apparently, some new ideas are needed. A special challenge to the theory of nonequilibrium systems, mentioned in the previous section, is to explain the range of slow relaxation rates observed in amorphous materials. Presumably, these rates are associated with metastable or, perhaps, weakly unstable states of such systems; but only little progress has been made so far in understanding the kinetics of this extremely important class of materials.

CONCLUDING REMARKS

As should be evident from the preceding discussion, the boundary between physics and materials science or engineering has never been sharply defined and is becoming even less distinct in the course of modern developments. The principal remaining distinction is one's point of view. The physicist, ideally at least, sees intellectual challenge in this area because novel materials pose qualitatively new conceptual questions or because familiar phenomena are recognized as specially tractable cases of far more general problems. The question is irresistible: Might a deep understanding of dendritic solidification provide some clue to mechanisms for cell differentiation or galaxy formation? The materials scientist, on the other hand, often seeks the best answers to specific complex problems and may have to substitute empirical knowledge for fundamental understanding in order to find them. The physicist may be unrealistic and the materials scientist shortsighted, but they are converging on identical scientific questions. Moreover, because of the remarkable advances in the tools available to these investigators, answers to many of those questions suddenly seem accessible.

5

Geophysics

INTRODUCTION

Geophysics is at the interface between physics and the geological sciences. It concerns the application of physics to understanding the structure and dynamics of the Earth and other bodies in the solar system. Geophysics includes meteorology (atmospheric science) and oceanography as well as solid-earth geophysics. Geophysics has developed as a discipline associated with the earth sciences more closely than with other fields of physics. Geophysics research, however, is firmly based on basic physical principles and presents complex problems that push the frontiers of fundamental physics; an outstanding example is the basic theory of fluid turbulence, which is the subject of intense interdisciplinary research at major geophysical laboratories. The concept of fractals and the renormalization group approach as applied to scale-invariant phenomena is applicable to many geophysical problems. Measurements of the inaccessible dynamic and static variables of geophysics present a continuing challenge that depends on new physics research for development of new physical methods; major experimental advances in geophysics since ancient times have been associated with the introduction of new or improved physical methods. Notable examples in recent years include deep seismic profiling, side-scanning studies of bathymetry, mass-spectrometric studies of isotopes, remote sensing from satellites, measurements of the turbulent spectra in the oceans and the atmosphere, and measurements of rock properties at ultrahigh pressures.

Until 1967 solid-earth geophysics included primarily the study of seismology, gravity, and magnetics. Geodesy provided the basis for map making and the orbital mechanics of the Earth. In 1967 the hypothesis of plate tectonics was proposed, and its subsequent acceptance provided a unifying basis for understanding a wide range of geological phenomena. In the past 15 years a rapidly advancing understanding of most geological phenomena has evolved: these include the worldwide distribution of seismicity, volcanism, and mountain building. Geophysics has made a substantial contribution to these advances.

The atmosphere and oceans form an interlinked system. The transport of gases, heat, moisture, particulate material, and momentum across the air-sea interface influences the evolution of climate, weather prediction, and the distribution of trace chemicals and human-generated pollutants. On both sides of this interface lie turbulent boundary layers. The interface itself is a convoluted surface, with breaking surface waves that inject bubbles into the sea below. Recent advances in observational oceanography and boundary-layer meteorology have led to substantial improvements in our understanding of small-scale mechanisms. For problems of global scale, we are coming to rely on remote sensing, particularly from orbiting satellites, to provide air-sea transfer maps.

Studies in geophysics have a wide range of practical applications. They provide the basis for the discovery and recovery of fossil fuels. Geophysical phenomena are also important to nuclear and geothermal energy. Geophysics provides the basis for understanding the major environmental hazards that are due to hurricanes, tornadoes, earthquakes, and volcanic eruptions. Weather, geodetic, seismic, gravity, and magnetic studies have a wide range of applications in terms of national security. Some of these applications are summarized here, and more appear in Chapter 11 of this volume.

Space missions have provided a wealth of data with a fundamental geophysical importance. Observations of the Earth have upgraded the data base in oceanography, meteorology, geodesy, gravity, and magnetics. Missions to the Moon and the planets have provided the basis for comparative planetology. Rocks returned from the Moon provide invaluable data on the early evolution of the solar system. Theories for the evolution of the Earth must be consistent with the data recently obtained from Venus.

A variety of experimental and theoretical studies in solid-state physics have direct applicability to geophysics. The complex nature of geological problems is ideally suited to such concepts as fractals and chaos; making large computers available to implement these mathematical models is also a high priority.

SCIENTIFIC BACKGROUND

Plate Tectonics

During the past 20 years a scientific revolution has occurred in geology; geophysics has made a substantial contribution to this revolution. Twenty years ago the worldwide distribution of mountain building, earthquakes, and volcanism was not understood. There was no systematic basis for predicting the location of fossil fuels and minerals. The hypothesis of plate tectonics changed all this. In the period from 1967 to 1970 this hypothesis first was proposed and then became generally accepted by geologists, geophysicists, and geochemists. Plate tectonics provides a natural explanation for continental drift; a small minority of earth scientists has given qualitative arguments for continental drift since the pioneering work of Arthur Wegener in the 1920s. Although the discovery of plate tectonics has been compared by some with the discovery of quantum mechanics, it is probably more appropriately compared with the discoveries of Newton.

Undoubtedly any reader of this volume is familiar with the basic axioms of plate tectonics: that the surface of the Earth is broken into a series of essentially rigid plates that are in motion with respect to one another with velocities of 1-10 cm/year, that new ocean floor is created at midocean ridges and old ocean floor is recycled into the interior of the Earth at ocean trenches, that the basaltic rocks of the ocean floor have a young average age of about 5×10^7 years because of the continuous recycling of the sea floor into the mantle, that the buoyancy of the thick continental crust prevents the continents from participating directly in the plate tectonic cycle (thus, the mean age of the continental crust is about 1.5×10^9 years), and that a substantial fraction of mountain building, seismicity, and volcanism occurs at the boundaries between plates.

The Earth as a Thermodynamic Engine

It has long been recognized that the propagation of shear waves through the Earth's mantle requires that the mantle be solid; yet the relative motion of the surface plates implies a fluidlike behavior. This apparent contradiction can be explained in terms of accepted solid-state creep processes; thermally activated creep, which is due to the migration of vacancies and dislocations, leads to a fluid rheology. Laboratory studies of the dominant mantle mineral olivine have shown that thermally activated creep processes yield reasonable mantle rheologies. Direct observational evidence for the viscosity of the mantle comes from studies of postglacial rebound. Data on dated elevated beach terraces have been inverted to give the viscosity as a function of depth in the mantle. This inversion gives a near-uniform mantle viscosity of 10^{22} poise (glacier ice has a viscosity of 10^{14} poise).

Once a fluid behavior for the mantle is accepted, plate tectonics is easily explained in terms of thermal convection. The surface plates are the thermal boundary layers of mantle convection cells. The cold rock of the oceanic lithosphere is denser than the mantle rock at depths when it is adiabatically compressed to the same pressure. There is a large downward (negative) bouyancy force on the sinking plates. The plates act as stress guides, transmitting the gravitational body force to the surface plates. This process is referred to as trench pull. Another major force on the surface plates is gravitational sliding off oceanic ridges. This force is entirely equivalent to a horizontal pressure gradient since the elevation of the ridge acts as a hydraulic head. This process is referred to as ridge push. Trench pull and ridge push are the dominant driving mechanisms of plate tectonics.

There are two primary energy sources that drive mantle convection and plate tectonics. The first is the heat released by the decay of the radioactive isotopes of uranium, thorium, and potassium. The second is the secular cooling of the Earth. Because the viscosity of the mantle has a strong temperature dependence there is a feedback between the rate of secular cooling and the efficiency of the convective heat transport mechanism. Calculations of the secular cooling contribution give values between about 15 and 50 percent. This result has important implications with respect to the early evolution of the Earth and the distribution of elements within its interior. The energy associated with seismicity, volcanism, and mountain building is about 0.1 percent of the energy available from radioactive decay and secular cooling.

Continental Deformation

The structure of the ocean basins can be largely explained by plate tectonics. However, the deformation of the continental crust is more complex. The structure of the Andes in South America can be attributed to the descent of the ocean plate along that continent's western margin. But the deformation of the western United States is much more complex. In general terms, the San Andreas Fault in California accommodates a fraction of the relative motion between the Pacific and the North American plates. However, the entire United States west of the Colorado front represents a broad zone of deformation accommodating the motion between the two plates.

The most spectacular mountain belt in the world is the Himalayas. This belt and the broad zone of deformation that extends throughout much of China is the result of a continental collision between the Indian and Eurasian plates. Broad zones of deformation are found in continents but not in the oceans. Continental plates appear to deform more easily than oceanic plates. An important area for future studies is the rheology of the continental crust. Mechanisms such as pressure solution are likely to be important and are poorly understood. The physical processes associated with folding and faulting are also poorly understood.

An important source of data concerning deformation in the continental crust is the Consortium for Continental Reflection Profiling (COCORP) project. This project, supported by the National Science Foundation, involves a full-time seismic reflection crew that utilizes oil exploration techniques to study the fundamental processes of crustal deformation. One result of this study is a better understanding of the formation of the southern Appalachians when the proto-Atlantic Ocean closed about 400 million years ago, resulting in a continental collision. A thin sheet of proto-Africa was thrust some 300 km over proto-North America along a low-angle fault. Another study has shown that the major mountain ranges of the overthrust belt of the western United States result from horizontal displacements along faults with a near-constant dip of about 30 degrees.

GEOCHEMICAL RESERVOIRS

Studies of the chemical evolution of the Earth are an important new frontier of research in geophysics. These studies involve an integrated approach to the evolution of the atmosphere, the oceans, and the solid earth. They emphasize the use of isotopic systems to monitor the evolution of the Earth. The Earth is divided into a series of reservoirs. In addition to the atmospheric and oceanic reservoirs, there are either three or four solid-earth reservoirs. These are the continental crust, the Earth's core, and either one or two mantle reservoirs. Those scientists who favor layered mantle convection argue that the chemical characteristics of the upper and lower parts of the mantle can be substantially different. They further argue that the lower mantle is the source of the anomalous chemical characteristics of the intraplate and hot spot volcanism (e.g., Hawaii and Iceland). Normal midocean-ridge basalts are extracted from the upper-mantle reservoir, which is depleted in many incompatible elements. The complementary enriched reservoir is the continental crust.

Important questions concern the formation of the reservoirs and the transport of material between them. There is considerable evidence supporting the very early formation of the core, oceans, and atmosphere. Although the core formed early, it is possible that there is transport from the core to the mantle. As the Earth cools the solid inner core is expected to increase in size. It is likely that the light alloying elements in the outer core come out of solution and rise to the core-mantle boundary. Some geophysicists argue that the motion induced by the rise of these exsolved elements drives the dynamo that creates the Earth's magnetic field. But the details of the dynamo mechanism remain one of the major unresolved problems in geophysics.

Radioactive dating of continental rocks shows that a substantial fraction of the continents have an age between 2 billion and 3 billion years and that the continents have continued to form until the present. There is considerable controversy concerning the recycling of continental crustal material into the mantle. Some scientists argue that substantial quantities of sediments are entrained in the oceanic crust when it is subducted into the mantle at ocean trenches.

The continental crust is a strongly enriched chemical reservoir that contains a substantial fraction of the incompatible elements in the Earth. Studies of isotopic distributions indicate that the mean age of the continents is 2 ± 0.5 billion years. Volcanic processes at island arcs transport strongly fractionated magmas into the continental crust. There is strong evidence that the depleted, stratified oceanic plate is mixed back into a nearly homogeneous (upper) mantle reservoir by vigorous mantle convection. This mixing must be substantially completed in less than 500 million years. However, the continental crustal reservoir is further differentiated and stratified by a variety of magmatic and hydraulic (particularly hydrothermal) processes. These processes lead to the formation of economic mineral deposits and are currently the subject of extensive study.

The oceans play an important role in the cycling process. Rivers and erosion provide a source of many elements to the oceans. Sedimentary processes return many of these elements to the seafloor. Studies from submersibles have demonstrated the importance of hydrothermal processes on ocean ridges. These processes transport elements from the oceans to the oceanic crust and from the oceanic crust to the oceans. The atmosphere plays a similar role for gases.

Isotopic determinations using mass spectrometry provide many of the relevant data concerning geochemical cycles. Extreme precision is required to study radioactive systems such as neodymium-samarium and lutetium-hafnium. Future developments in mass spectrometry are expected to make a number of other isotopic systems accessible to study.

The Atmosphere and Oceans

What makes the geophysical fluid dynamics of the atmosphere and oceans challenging is the factor-of-10^{10} difference between the scales of motions of planetary scale and the motions of smallest scale, where molecular diffusion is important. Thus, a theory or computer simulation of the weather must incorporate the cumulative effect of all the smaller-scale fluid dynamics: internal waves, fronts, two- and three-dimensional turbulence, and convective clouds. Intense studies of these intermediate scales of motion are being pursued with applications to severe storms, cloud modeling, and frontal dynamics.

A simulation of climatic change must accurately account for the many years of weather, whatever its cumulative effect may be. A theory of ocean circulation, on the other hand, must cope with the vastly slower response of the ocean to changes in atmospheric winds and heating. It must account also for the differing behaviors of salinity and temperature, both of which influence the fluid buoyancy. The worldwide disruption of weather by the 1982-1983 El Niño event in the tropical Pacific Ocean shows the powerful interactive nature of oceans and atmosphere. Analytical theories for the propagation of geostrophic waves have successfully described several of the links in the sequence of tropical and global change. Future studies will require the most modern computational facilities.

Beyond these short-term events we turn to the global effect of increasing carbon dioxide in the atmosphere. The prediction of climate change over the next half century relies on complex fluid dynamical modeling of the general circulation and its heat and moisture balances. These problems involve ocean-atmosphere interactions and the input of biota. Remote sensing from space vehicles is a promising source of new data.

Turbulence remains one of the major unsolved problems of physics. The geophysical aspects of turbulence are an interesting special case. The atmosphere and oceans are the greatest natural laboratory for turbulence. It is not an exaggeration to say that weather prediction, the understanding of the general circulation of the atmosphere-ocean system, and the evolving states of climate are all problems of fluid dynamical turbulence.

There are two distinct kinds of geophysical turbulence: small-scale turbulence, which is generally isotropic, and large-scale turbulence, which is affected by stratification and planetary rotation. In fact, when these geophysical constraints are added, the turbulence problem is related to both wave propagation and flow stability.

Of particular interest is the manner in which turbulence and waves interact in systems with buoyancy and rotation. These chaotic motions interact with the general circulation of the fluid in a strong manner. A key question subject for current study is the way in which eddies and the mean circulation influence one another. We are now approaching a clear dynamical understanding of diverse phenomena, such as the Southern Oscillation, stratospheric sudden warming, quasi-biennial oscillations of the atmosphere, and the wind-driven gyres and deep thermodynamically driven circulation of the oceans. At smaller scales turbulent mixing of the oceans determines the gross structure of the temperature, salinity, and trace chemistry.

Changes of phase of water are at the heart of climate dynamics. Evaporation at the sea surface and recondensation in tropical cumulus clouds provide the principal force for the atmospheric circulation. The evaporation process is embedded in turbulent boundary layers both above and below the sea surface. The condensation of water vapor into clouds requires condensation nuclei, and there is a complex of surface chemistry and particle kinetics interactions with the fluid dynamics of the convecting cloud. Beyond the dramatic examples of these heat engines, such as hurricanes, the dynamics of tropical cloud clusters is a central problem in atmospheric dynamics. Parameterization of cloudiness and precipitation in numerical models of the circulation is a significant unsolved problem. Successful three-dimensional simulations of rudimentary, individual convecting clouds are now being carried out; they require the fastest available computers.

On land, the hydrologic cycle is a fluid-dynamical problem involving phase change and flow through a complex porous medium. As with the ocean, the groundwater and snow cover act as a memory for climate evolution and long-range weather predictions.

We are seeing rapid progress in the understanding of the general oceanic circulation, both the mechanical response to the stress exerted by the winds overhead and the thermodynamic response to the heat flux and the moisture flux between the air and the sea. These theories of the circulation, wave propagation, turbulent cascades, and the induction of mean circulation by eddy motions are laying the groundwork for the coupled model of the ocean-atmosphere system. The close interaction of theory, observation, and computer and laboratory experiment is characteristic of the work.

Comparative Planetology

THE MOON

The first manned landing on the Moon took place on July 20, 1969, and introduced a decade of comparative planetology. Studies of the physical properties and chemistry of the rocks returned from the Moon provided a wealth of information on the early solar system. Chemical studies showed that the lunar maria are composed of basaltic rocks similar in major element chemistry to basaltic rocks of the Earth's oceanic crust. Radiometric dating of these rocks gave crystallization ages of 3.16 billion to 3.9 billion years. Rocks returned from the lunar highlands had a much more complex chemical history and had been extensively altered by meteorite bombardment. Dating of these rocks gave ages of about 4.5 billion years, close to the estimated age of the solar system.

Although we know a great deal more about the Moon today than we did 15 years ago, the question of its origin remains unresolved. There are three hypotheses: (1) that the Moon was originally a separate planet and was captured by the Earth, (2) that the Moon was originally part of the Earth and that the Earth fissioned into two parts, and (3) that the Earth and the Moon formed as a binary planet. Each hypothesis has strong advocates and substantial difficulties.

MARS

Detailed photographs of Mars returned by the Mariner and Viking spacecraft have provided a wealth of information. It is clear that Mars bears little resemblance to either the Earth or the Moon. The most striking global feature of the Martian surface is its hemispheric asymmetry. Much of the southern hemisphere of Mars is covered by densely cratered terrain that is probably the remnant of the postaccretionary surface of the planet, whereas most of the northern hemisphere is made up of high-standing lightly cratered plains that are younger and probably of volcanic origin.

There are shield volcanoes on Mars that dwarf the largest volcanic structures on the Earth. The largest of the Martian volcanoes are in the northern hemisphere in an area of elevated terrain known as the Tharsis uplift. Four of these volcanoes are about 21 km above the Mars reference level; the most spectacular of these is Olympus Mons, with a mean diameter of 600 km and a summit caldera 80 km in diameter.

Although the Viking landers provided some information on the composition of the surface rocks, hypotheses concerning the evolution of the body must await the return of a variety of surface samples to the Earth. Recent studies have associated several classes of meteorites with a lunar and Martian origin. Nevertheless, a Mars sample-return mission, either manned or unmanned, must be one of the highest-priority space missions in the next 25 years.

VENUS

In terms of comparative planetology, Venus closely resembles the Earth; its overall mass and mean density are quite similar to those of our planet. However, the thick cloud cover prevents direct observations of its surface. Earth-based radar studies and the radar altimeter on the Pioneer Venus orbiter have provided detailed data on the surface topography of Venus. The planet is remarkably smooth: 64 percent of the surface is a plains province with elevation differences of 2 km or less; highland areas stand as much as 10 km above the plains, but they constitute only about 5 percent of the surface; lowlands are 2-3 km below the plains and occupy the remaining 31 percent of the surface. Most of the highlands are concentrated in two main continent-like areas: Ishtar Terra, the size of Australia, in the northern hemisphere, and Aphrodite Terra, about the size of Africa, near the equator. The summit of Maxwell Montes, a possible volcano in Ishtar Terra, is the highest point on Venus, about 11 km above the mean reference surface. The lowest point on Venus is in a trench or rift valley about 3 km below the reference surface. Clearly we must learn much more about Venus. However, the high surface temperature and corrosive atmosphere make accurate surface measurements extremely difficult to obtain.

APPLICATIONS

Hazards

SEISMICITY

Earthquakes present a serious hazard to life and property, particularly in the contiguous western United States and Alaska. At present there is no way to predict seismic events, but plate tectonics provides a rational basis for hazard evaluation on a limited basis. The recurrence of large earthquakes (greater than magnitude eight) on the San Andreas Fault is well established. The recurrence period is about 150-200 years. Since the last great earthquake in the Los Angeles area occurred in 1857 (49 years before the great earthquake in San Francisco), it is generally accepted that the next great earthquake will occur in the near future in Southern California. Clearly, precautionary steps should be carried out. Emergency preparations are fully justified. The greatest hazard to life would result from the failure of dams. Such a catastrophe came close to realization in the 1972 San Fernando earthquake. Based on experience in other parts of the world, the risk of landslides in general and in particular landslides into the major reservoirs in the San Gabriel Mountains constitute a major peril.

Severe hazards are also posed by smaller earthquakes on the San Andreas system. The 1972 San Fernando earthquake occurred on a fault whose existence was not even known before the earthquake. Similar faults pervade the heavily populated Los Angeles basin. The seismic risk extends throughout the western United States. One example is the Wasatch Fault, which is a documented, highly active fault in Utah. The fault scarp in Salt Lake City is the site of extensive apartment and real-estate developments because of the excellent views that the scarp provides. Although seismic activity in the eastern United States is much less frequent, large earthquakes do occur. The 1812 earthquake in New Madrid, Missouri, may have been the largest shock to have occurred in the United States in the past 200 years. The origin of these intraplate earthquakes is poorly understood.

VOLCANIC ERUPTIONS

The eruption of Mount Saint Helens provided dramatic evidence of the threat to life posed by the active volcanoes in the western United States. It was the latest in a series of major explosive eruptions that have occurred in the Cascade volcanic chain. There is also a risk of massive eruptions in other parts of the western United States. Fortunately, the prediction of volcanic eruptions using seismic and geodetic techniques has been quite successful.

Energy

FOSSIL FUELS

The largest demand for trained geophysicists comes from the petroleum industry. Seismic reflection profiling is the dominant technique for discovering petroleum and natural-gas reserves. Sophisticated data-processing techniques using the largest available computers allow reservoirs to be identified directly from reflection records. This is a highly proprietary, rapidly developing field.

Another major application area involves down-hole logging. Geophysical observations in wells provide a wealth of information necessary for the evaluation of reservoir properties. A major source of petroleum in the short term results from secondary and tertiary recovery techniques. These techniques have substantially increased the fraction of the petroleum that can be recovered from a reservoir. The physics of multicomponent flows in porous media is poorly understood, and its elucidation requires a high priority.

Although we are now enjoying a period in which the supply of petroleum exceeds the demand, it is only a matter of time until the available reserves are exhausted. Recent oil company explorations in the offshore Atlantic, eastern Gulf, and offshore North Slope failed to find large hydrocarbon deposits and are evidence that few new major oil fields will be found. Nevertheless, fossil fuels will be the major source of energy for the foreseeable future. There are immense reserves of tar sands, oil shales, and coal. Recovery is primarily an economic and environmental problem, but basic research on new recovery techniques obviously should receive a high priority.

Although the environmental hazards associated with nuclear energy receive a large fraction of the attention of environmentalists, there are certainly environmental hazards associated with the combustion of fossil fuels. One that is receiving considerable attention is acid rain. Another is the climatic effects associated with the increased production of carbon dioxide. The geochemical cycle of carbon dioxide is poorly understood, and its study also should receive a high priority.

NUCLEAR ENERGY

Two of the major problems affecting the nuclear-energy industry are directly related to geophysics. The first is the seismic hazard to nuclear power plants, and the second is the problem of waste disposal. A primary factor in the escalating cost of nuclear power plants is the construction codes dictated by seismic-risk assessment. Although a rational basis for risk assessment exists in plate boundary areas such as California and, to a lesser extent, the rest of the western United States, there is little or no basis for risk assessment in the eastern United States. As we mentioned above, although the frequency of seismicity in the eastern United States is low, large earthquakes are known to occur. The 1812 earthquake in New Madrid, Missouri, had a magnitude comparable with that of the 1906 San Francisco earthquake. The possibility that another earthquake of similar magnitude could occur throughout much of the eastern United States has led to extremely conservative building codes that greatly escalate building costs of nuclear power plants. A better understanding of the seismic hazard within plate interiors could make nuclear energy economically feasible, at least in parts of the United States.

The problem of nuclear-waste disposal is now receiving a high priority. A variety of sites have been proposed for nuclear-waste repositories. The long-term integrity of the various sites poses a number of geophysical questions, including the effects of heat production on a site, the mobility of groundwater, and the basic geological stability of the various sites.

GEOTHERMAL ENERGY

On a local basis geothermal reservoirs can provide clean, cheap power. The primary example in the United States is the Geysers field north of San Francisco. At present, sufficient power is being generated from this field to meet the power requirements of the city of San Francisco. The Geysers field produces clean dry steam that can be run directly through turbines.

Because of the proprietary nature of the search for geothermal reservoirs, few data on reserves are available. It is doubtful that other fields as productive as the Geysers exist in the United States. The Imperial Valley of California contains large reserves of geothermal energy, but the corrosiveness of the hot brines presents severe technical problems that have not been solved.

The fundamental physical processes associated with geothermal reservoirs are poorly understood. Reservoirs are treated as nonrenewable gas fields. It is generally accepted that the high temperatures required for steam are associated with solidifying magma bodies. If geological conditions restrict groundwater circulations, the groundwater may be heated sufficiently to form reservoirs of wet or dry steam.

It has been proposed that heat be extracted from hot, dry rock. Cold water would be injected into one well, and the heated water would be extracted from a second well. The circulation path would be continuously altered so circulation would take place through hot rock. Although the process is feasible in principle, the technical problems may be impossible to overcome.

It is highly unlikely that geothermal energy will be a major source for this nation's energy supplies. However, on a local basis, geothermal reservoirs can provide significant amounts of cheap energy.

Data Acquisition from Space

Data acquired from satellites have provided a wealth of geophysical data. Satellite tracks can be inverted to obtain the Earth's gravitational field. An accurate gravitational field is required to predict satellite orbits. Modern geodetic observations from space require an orbit location of better than a meter. The GEOS-3 and SEASAT satellites mapped the sea-surface height to an accuracy of better than 10 cm. This not only provided an accurate geoid map over the oceans but also provided important data on ocean currents and tides. MAGSAT and other satellites have provided an accurate map of the Earth's magnetic field.

A series of LANDSAT missions has provided a wide range of spectral data on the Earth's surface. These data can be used to produce geological maps of areas of sparse vegetation. An important new data base is being collected by the use of side-scanning radar to obtain maps of the Earth's surface.

National Security

Detection of underground nuclear explosions provided the basis for modern seismology in the 1950s. Today, the most sophisticated seismic arrays are dedicated to this purpose. Research on the behavior of substances subjected to high pressure, carried out to elucidate the state of the Earth's interior, is used to increase our understanding of the effects of a nuclear explosion.

Geophysical observations have long been of fundamental importance to national security. Topographic maps provide the basis for military operations and have secret classification in many parts of the world. Detailed swath mapping of bathymetry was highly classified for a long period. Mapping of the Earth's gravitational field provides the basis for plotting missile trajectories and navigation, and maps of the Earth's magnetic field also provide an important aid in navigation.

FUTURE DIRECTIONS OF RESEARCH

Seismic Studies of the Continents

Seismic studies of the continents complement the seismic reflection studies being carried out by the petroleum industry. Seismic studies in sedimentary basins are the primary basis for finding new petroleum resources. COCORP has carried out reflection studies for scientific purposes for the past 10 years. Major scientific discoveries have been made, and potential oil and gas provinces have been discovered. Many other countries have undertaken similar projects.

Despite efforts made over the past 10 years the structure of the continents remain largely unexplored. Modern tomographic analysis techniques are enhancing the effectiveness of seismic reflection and refraction studies. The major benefits in terms of the discovery of resources, the mitigation of seismic and volcanic hazards, and major scientific discoveries that can be easily envisaged suggest the need for strong support for these studies.

Deep Drilling in the Continents

The Soviet Union is currently drilling a continental well that has reached a depth of 12 km. The well has been drilled entirely through crystalline rock and is 5 km deeper than the deepest previous well, which was drilled entirely through sedimentary rock. Preliminary information indicates that the Soviet Union has made a number of important scientific discoveries concerning heat flow, the presence of mineralized water, and the type of rock present. Deep drilling in selected areas, such as that proposed by the corporation for the Deep Observation and Sampling of the Earth's Continental Crust (DOSECC), should provide exciting new discoveries.

Global Digital Seismic Array

Studies inverting the travel times of both body and surface waves generated by earthquakes show great promise for determining the structure of the interior of the Earth. These studies utilize the latest developments in tomography. In order to obtain an accurate, small-scale distribution of seismic velocities within the Earth a global distribution of digital seismic stations is required. The technology that is available is not suitable for use at deep ocean sites, thus leaving holes in the network to be resolved through future research. Deep ocean seismometers need to be developed. Such a global network would also be of great value in the detection of underground nuclear tests.

Large-Scale Computing Facilities

There is a multitude of demands for large-scale computing facilities in geophysics. They are essential for performing calculations of the circulations of the oceans and atmosphere. Inversion of seismic reflection profiles and data from the global seismic network requires the use of the largest available computers.

Studies of Crustal and Mantle Materials Under High Pressure

In order to understand processes that are occurring within the Earth it is necessary to know the physical properties of the relevant minerals. One of the most successful devices developed recently is the diamond pressure cell. Combined with the high level of radiation available from synchrotrons, this device can produce a wide range of important results on phase diagrams and reaction kinetics. Some studies of physical properties require larger sample sizes than can be accommodated in diamond cells. An example is creep experiments to determine the rheology of relevant minerals at high pressures. A variety of experimental facilities for high-pressure studies is needed.

Remote Sensing from Space

Remote sensing of the Earth from space provides a variety of data concerning fundamental geophysical processes. Optical and microwave studies provide a range of information on meteorological and ocean processes. With future developments direct observation of precipitation may be possible. An important objective is to improve the prediction and tracking of hurricanes and tornadoes.

The use of the Global Positioning System (GPS) is certain to revolutionize geodesy: absolute positions will be routinely determined with centimeter accuracy at little cost. Such observations will greatly improve our knowledge of seismic and other tectonic processes. The GPS will be extremely useful for many commercial fields involving surveying and navigation.

Other Geophysical Data Sets

Understanding fundamental geophysical problems requires extensive data sets. This is particularly true in meteorology and oceanography. Extensive and detailed studies of velocity fields and spectra are required. Measurements of temperature, salinity, and moisture content are also necessary. In solid-earth geophysics, electromagnetic studies show considerable promise toward helping us to determine the temperature, fluid content, and composition of the crust and upper mantle. Heat-flow measurements provide information on mantle convection and heat-transfer processes.

6

Computational Physics

INTRODUCTION

Computational physics extends theoretical physics beyond the limitations of analytic techniques. This extension has become essential to the advance of many different subfields of physics as systems of interest have become more complex, moving from a few degrees of freedom to many degrees of freedom and from linear to nonlinear problems.

Computational physics is not a subfield of physics, as is, e.g., elementary-particle physics or condensed-matter physics. Nor is computational physics a third mode of physics, parallel to experimental physics and theoretical physics—although the misconception is sometimes adduced. Computational physics, since it does not adduce new experimental facts about the physical universe, must lie wholly within the sphere of theoretical physics. It is distinct from, and should not be confused with, the essential uses of modern computers in physics experimentation.

As today's experimentalist relies on the power of the computer to extend his or her abilities to gather and process data, many theorists of today (and, we think, most theorists of tomorrow) are coming to rely on computational power to extend the mind's ability to imagine theoretical models and to discern the often complex and subtle consequences and predictions of those models.

Computational physics is an interface subject not only to fields outside physics but also (and perhaps more importantly) among the various subfields of theoretical physics, among them particle and field theory, condensed-matter theory, plasmas, fluids, and atomic physics. In this chapter, our agenda is threefold:

• First, we elaborate on the different types of complexity which lead theoretical physics naturally and inevitably into the computational realm; we start in general terms and then give specific examples from a number of different subfields of physics.

• Second, we discuss the distinctive nature of computational physics as it affects the lives and perspectives of its physicist practitioners.

• Third, we try to extract sensible policy implications: What needs to be done to ensure that computational physics, and the information science revolution generally, is optimally harnessed in the service of the advance of our knowledge of the physical universe?

A number of recent panel and committee reports have addressed these and related issues and can usefully be read in parallel with this one. The 1981 *Report by the Subcommittee on Computational Facilities for Theoretical Research* of the National Science Foundation's Advisory Committee for Physics (Press report) is similar in scope to this chapter; its descriptive material has been drawn on heavily for this chapter. The 1982 *Report of the Panel on Large Scale Computing in Science and Engineering* (Lax report) looks at computation not only in physics but also in other disciplines. In the context of a particular funding agency, the National Science Foundation, the 1983 report *A National Computing Environment for Academic Research* (Bardon report) makes detailed, and to our minds cogent, recommendations.

THEORETICAL INVESTIGATION OF COMPLEX SYSTEMS

As the fields of physics advance, theory naturally evolves from a stage where the most important problems can be solved analytically to a stage where numerical and other computational techniques (such as computer symbolic manipulation) become essential. At any one time, different subfields are in different stages of this process. However, one can identify—even across the boundaries of different subfields— common features that drive this evolution. In general terms, one moves from a few degrees of freedom to many degrees of freedom and from the linear or linearized to the intrinsically nonlinear.

Complexity sometimes emerges in moving from ordinary differential equations to partial differential equations for the solution of interesting problems. This has happened, long since, in the fields of fluid and plasma physics, and for the Schrödinger equation in noncentral or many-body problems, for radiative transfer problems in astrophysics, and in other areas. Similarly, in areas such as nuclear reaction theory, many coupled integrodifferential equations naturally arise and must be solved self-consistently.

Complexity also increases dramatically when one moves from simple or one-dimensional models of physical processes to realistic simulations. One sees this in research on the phases of fluids and solids and lattice models, in plasma simulations, in many-body calculations of galactic dynamics, and in quantum field theory.

Complexity can come from the impetus to move from low order to high order in any of several mathematical expansion schemes. This is the case in quantum electrodynamics and quantum chromodynamics; in high-temperature and other expansions for liquids, solids, and lattice systems; and in high-partial-wave expansions for nuclear reactions. When the expansion techniques are algebraic, one moves from hand algebra to symbolic manipulation on the computer.

Complexity arises in moving from scalar systems to vector or tensor systems and from linear systems to nonlinear systems. A striking example of this is gravitation theory (general relativity), whose partial differential equations are both tensor and highly nonlinear. Another example is the physics of many-component mixtures of fluids. In all these cases, one sees that complexity arises not from an injudicious choice of problems but inevitably as theory advances. Surveying the scope of subfields of physics, as we shall now do at least in part, one finds over and over again that (1) there are important problems whose solutions must be found by computational techniques and (2) even the best investigators lack sufficient resources for accomplishing even the tasks already at hand.

ELEMENTARY-PARTICLE PHYSICS

In quantum electrodynamics it is possible to extend perturbative solutions to high order, and large-scale computing has played an essential role. The calculations of the sixth-order and eighth-order magnetic moment of the electron, and their impressive agreement with experiment to 1 part in 10^8, have provided one of the most accurate verifications of a fundamental theory in all of physics.

It has now become necessary to carry out equally complex calculations for quark theory (quantum chromodynamics) in order to make realistic comparisons of theory with experiments. The algebra of these calculations, in particular, is extremely complex, and computer-based algebra programs such as SMP and MACSYMA have become crucial. In addition, the integrals can usually be found only through numerical Monte Carlo integrations.

In continuum quantum field theory, there is no known method for numerical approximation of the nonperturbative regime. However, the formulation of lattice versions of currently viable gauge theories has led to substantial new insights into the theory of quark confinement and has also opened up the application of statistical-mechanical high-temperature expansion techniques to study the behavior of gauge theories for large values of the gauge coupling constant. Monte Carlo calculations for the pure gauge theory have already had a substantial impact on elementary-particle theory. However, such calculations are just a beginning. Substantial increases in computing power are needed to permit more realistic calculations that use lattice sizes large enough to show independence of the lattice itself, to incorporate quarks in a realistic way into the calculations, and to use the physically relevant color SU(3) group in place of the simpler models mostly studied to date.

Efforts are under way in several laboratories to meet the need for enormous computing power for quantum chromodynamics problems by constructing special-purpose parallel processors dedicated to this problem. Although such computers are not yet operational, their published designs indicate (special) capabilities that dwarf currently available general-purpose vector computers. This fruitful general approach is likely to stimulate theorists to invent special-purpose computers and to seek their construction.

One of the most fundamental and baffling problems facing elementary-particle theorists is to understand dynamical breaking of chiral symmetry, namely, a breaking that does not require Higgs fields. This occurs in strong interactions and is an essential part of a major class of unified theories of weak and electromagnetic interactions. It occurs only in strongly interacting field theories, so it cannot be studied perturbatively. To date, no lattice approximation has been found that maintains continuous chiral symmetry. An important research effort at the moment is the search for a framework that permits the numerical study of dynamical symmetry breaking. If such a framework is found, it is certain that the computing requirements to study this problem (which involves both fermions and gauge fields) will be immense.

STATISTICAL MECHANICS

An important chapter in the history of the theory of critical phenomena in matter (magnets and fluids, for example) was the use of high-temperature expansions, to very high order, to compute critical exponents. Large-scale computing was required to push the series beyond the level computable by hand; this was done to check the consistency of extrapolations to the critical temperature from very high temperatures. This work established the failure of mean field theories of critical behavior and paved the way for the renormalization group approach to describing critical behavior. This effort continues both through higher-order calculations for simple systems and through new, more complex, applications. Some new applications (e.g., multicritical points) involve expansions in several variables that require considerable computing power.

Another fundamental result in computational statistical mechanics was the calculation of the equation of state of the hard-sphere liquid using molecular-dynamics methods. Many detailed studies of the liquid state have been made by using perturbations of this model. Molecular-dynamics and Monte Carlo methods are now being used to investigate increasingly complex liquid systems, such as the simulation of liquid water.

CONDENSED-MATTER THEORY

New computational methods have been developed recently that allow, for the first time, accurate solutions of the complex many-body problems of condensed-matter theory. Many of these developments have been sparked by the increasingly urgent demands of experimentalists for theoretical interpretation of data obtained with new, sophisticated techniques. Many such problems are in the area of real materials and effects where techniques are now available to calculate answers to many vital questions in condensed-matter science. These include electronic structure determinations of complex bulk solids, polymers and biomaterials, impurities and vacancy defects, superfine particles, disordered solids, and amorphous materials such as modulated (superlattice) structures and sandwiches. Several of these problems focus on the unexplored regime that lies between molecules and solids, which is important for the future development of electronic devices.

Problems in which increased computational capacity and innovative numerical techniques are having wide impact include: static and dynamic structure of quantum liquids and their surfaces, the static and dynamic structure of classical liquids and their surfaces, the nature of homogeneous systems (the microscopic description of liquid-solid interfaces, for example), solitary excitations in extremely nonlinear systems, the vast area of nonequilibrium properties of solids and liquids (for example, the microscopic simulation of shock fronts in condensed matter), turbulence in both classical and quantum fluids, the simulations of lattice defects, and the electronic structure of bounded systems such as thin films, slabs, and absorbed layers.

To give some idea of the impact of increased computational capacity, it may be worth expanding on the last item by way of example. Only with the advent of the linearized, augmented plane-wave method did realistic self-consistent calculations of the electronic levels of thin metal films, including films with ordered overlayer absorbates, become possible. Even so, relatively modest systems strain the resources of most computers. Although further improvement in calculational schemes can be expected, these will come only from the understanding of extensive calculations of related systems, if the experience with bulk systems is any guide. Of greater interest is the construction of potential surfaces for molecules outside a transition metal surface, a project requiring a 10- to 100-fold improvement in computational capacity. Such surfaces can be the input for studying the sticking of atoms on surfaces, the absorption and disassociation of molecules, and the subsequent motion and reaction of atoms on the surfaces. All these have been the subject of primitive calculations, which could be improved if the requisite computational facilities were available. For the more distant future, there is the problem of the simulation of finite temperature surfaces, where the entropy aspect of reactions must be included.

ATOMIC AND MOLECULAR PHYSICS

Atomic and molecular science today is characterized by rapid advances in experimental technique, especially the ability to prepare and control a wide variety of atomic and molecular states. Highly ionized species, atoms in strong external fields, states with many electrons excited, states with dimensions approximating the macroscopic, and high-angular-momentum states are but a few examples. The properties of such states and their interactions play a central role in atomic physics research.

The availability of powerful computers has enabled theorists to make significant contributions to the rapid growth of atomic and molecular

SCIENTIFIC INTERFACES AND TECHNOLOGICAL APPLICATIONS

114

physics during the past 15 years. Old methods have been applied to more-complex problems, and new methods have been developed for the study of many of the processes that are now amenable to experimental study or are of interest to applied physicists. In the determination of electronic structure, calculations of wave functions for atoms and small molecules have progressed well beyond the Hartree-Fock level. However, present computing power and theoretical techniques are insufficient for accurate multiconfiguration calculations for heavy atoms in which relativistic effects are important. Such calculations will be required in the study of heavy-ion fusion and are needed, for example, for the analysis of the experiments searching for parity-violating atomic transitions. Further development of radiation physics and laser optics will require broader and more detailed studies of photon-atom interactions, often with highly ionized or perturbed atoms. Recent investigations of bremsstrahlung, Rayleigh scattering, Compton scattering, and the photoelectric effect have revealed interesting new phenomena that have been explored by only a few groups with extraordinary access to fast computers. At lower energies better calculations of photoionization will be necessary to interpret the wealth of new data generated with synchrotron light (for ground-state atoms or molecules) and with infrared or visible lasers (for highly excited states). With respect to larger systems, self-consistent field calculations can be carried out using the local density or local spin-density approximations for polyatomic molecules, including polymers and weakly bound clusters, and for molecules absorbed on surfaces. Calculations by better methods will facilitate the assessment of the accuracy of these approaches, and further applications of these methods should encourage greater collaboration among atomic physicists, quantum chemists, solid-state physicists, and biochemists.

In the study of atomic collisions, theory is now capable of verifying and augmenting experimental measurements on many processes in electron-atom collisions. There have been some notable successes in the theory of electron-molecule and ion-atom collisions at both high and low energies. Serious problems remain, particularly at intermediate energies and for collisions involving molecules in which electronic or vibrational excitation is important. Useful calculations on rearrangement collisions, energy transfer, excited-state reactions, and breakup processes will require new methods and increased computing power. The successful methods should be extended to treat collisions with atoms or molecules on surfaces. Many of the new diagnostic techniques for studying plasmas and solid surfaces involve atomic collisions, and more detailed calculations of the energy, angular distribution, and polarization of scattered particles will be needed if these techniques are to be utilized fully. Studies of electron-atom and atom-atom collisions in the presence of a laser field give information not otherwise obtainable. The calculations are necessarily difficult, however, and require extensive computational effort.

Monte Carlo techniques have been introduced into the study of the electronic structure and interactions of atoms and molecules within quantum, semiclassical, and purely classical theories. Simulations are also being used to relate the macroscopic behavior of ionized gases to the properties of the individual atoms and molecules. These simulations have led to significant improvements in transport theory and to a better understanding of swarm measurements of the reactions of atomic ions. However, further studies of energy exchange between molecular ions and neutral ions and molecules are needed. Better-designed simulations would be valuable in the exploration of the many-body effects that occur in dense gases, about which little is currently understood. For example, computer simulations of three-body recombination should help to clarify many of the mysteries concerning combustion at atmospheric pressure.

PLASMA PHYSICS

Theoretical plasma physics today includes many active subfields. Their common theme is the importance of collective processes, which dominate the interaction of ionized gases with electric and magnetic fields. Basic plasma theory addresses the fundamental natures of plasma turbulence, statistical mechanics, and magnetic-field topology. Equally important are the major applications of plasma theory to controlled fusion research, solar-system and magnetospheric plasmas, and astrophysical plasmas.

Areas of basic plasma physics theory that have shown rapid and exciting advances in the past few years include strong turbulence, soliton formation and dynamics, magnetic reconnection, plasma heating by intense beams, and single-species plasmas.

Numerical techniques for computational plasma physics can be grouped into two classes:

1. Techniques that follow the microscopic electrogmagnetic fields and particle-distribution functions are ideal for providing detailed information on the effect of turbulence and on the growth and saturation of strong plasma instabilities. However, they have the disadvantage of being computationally slow, so they cannot be used to model

long-time-scale phenomena. Approaches that describe the microscopic nonlinear physics include particle-in-cell computer simulations with up to 400,000 individual particles and codes that solve plasma kinetic equations. The latter involve partial differential equations that are functions of seven independent variables (three space dimensions, three velocity dimensions, and time). The limited capability of present computers places a strong limit on the number of space and velocity variables that can be used in practical kinetic-equation calculations.

2. For a macroscopic and long-time-scale description of the plasma, one uses a variety of fluid codes (which solve moment equations) or else hybrid codes that can average over fast time scales and short spatial scales. The state of the art in fluid computations is represented by the three-dimensional, resistive magnetohydrodynamic (MHD) calculations now being done on supercomputers. These require many hours of computing time per run. Hybrid codes are a more recent and rapidly evolving computational method and represent an area of major research effort at present. It seems likely that the next few years will see hybrid physics codes synthesized into comprehensive and realistic computational models. The computing requirements are expected to be immense.

NUCLEAR PHYSICS

Examples of recent, significant computational work within nuclear physics include large-basis shell-model calculations, hypernetted-chain calculations of nuclear matter binding energies with comparisons to Brueckner theory, the time-dependent Hartree-Fock theory of heavy-ion collisions, calculations of high-energy scattering with multiple-scattering corrections, few-body methods for three- and more-body processes, and fast coupled-channel reaction formalisms. New work in intermediate-energy physics has included the effects of virtual mesons and baryon resonances on nuclear structure and reactions, predictions of the ion-nucleon interaction from field theory, studies of the ion-nucleus optical potential, and derivations of properties of the nucleon-nucleon interaction from meson theory or quark models based on quantum chromodynamics.

Large-basis shell-model calculations continue to be important for understanding nuclear structure and reactions and as a microscopic model for developing and testing new approaches to treat finite nuclei. For example, the interacting boson approximation has had some remarkable successes in describing the complicated spectra of rotational and transitional nuclei, and it promises to play a major role in explaining the spectra of heavy nuclei far from stability and at intermediate excitation energies. The full exploration of this model and its relation to microscopic theories of nuclei requires extensive computing.

Statistic moment methods are a new approach to finding quantum averages of variables in a many-body system by calculating the traces of the relevant operators in large Hilbert spaces. There is some evidence that highly accurate calculations of nonaverage properties of nuclei can be made by using traces in superlarge spaces of relatively low powers (four or five) of the Hamiltonian. Even so, these calculations require many hours on the fastest computers available.

The coupled-channel reaction theory describes nuclear reactions by using a large number of coupled integrodifferential equations. It appears to be a fruitful approach for describing reactions that proceed through a number of different channels. The computers generally available are inadequate for serious calculations of reactions on intermediate to heavy nuclei.

The theory of few-body (three to five) reactions has been developed considerably in recent years. Several approaches now exist that provide well-posed sets of integral or differential equations for describing three- and four-body systems. Approximate and preliminary calculations have been successful almost beyond expectations, but access to the fastest existing computers is necessary for the further exploration of these methods in nuclear structure and reaction calculations.

A variety of approaches have been developed for the theoretical investigation of heavy-ion collisions, currently one of the two major areas of experimental nuclear research. These approaches deal with both the low-energy and the high-energy regions: (1) statistical multistep calculations using distorted-wave Born approximation matrix elements, (2) coupled-channel equations, (3) time-dependent Hartree-Fock methods, (4) cascade theory, and (5) hydrodynamic calculations. Although extensive calculations have been made that use some of these methods, the results point to the need for computationally more demanding investigations not only to exploit these methods but to test new theoretical ideas.

Cascade calculations using Monte Carlo methods provide virtually the only means for explaining the global features of reactions initiated by energetic hadrons. Gross features can be obtained from sampling a relatively small number of cascades, perhaps 5000, but recent experiments have probed rarer and therefore more significant events. Calculations of these events require large statistical samples, more than 100,000 cascades, and long computer runs.

PHYSICS OF FLUIDS

The frontier problems of fluid dynamics can be divided into two broad areas: basic physical understanding of nonlinear dynamics and the phenomenology of complex flows. Both kinds of problem challenge the capabilities of modern computers.

The three-dimensional character of many flows is essential to their proper understanding. Many important effects never appear in two-space dimensions. Among important three-dimensional problems, transition and turbulence involve the calculation of flows with a wide range of excited scales of motion. When the range of excited scales increases by only a factor of 2 (which occurs when the Reynolds number of the flow increases by about a factor of 2), the computer power necessary to calculate this flow increases by an order of magnitude.

The calculation of real fluid flows usually involves several other difficulties, including complicated geometries and topologies, multiphase flows, thin boundary and internal layers, interacting shocks, nonlinear instabilities, and material interfaces. All these problems are exacerbated when extra physics, such as chemical reactions, magnetohydrodynamic effects, and radiation physics, are included. Numerical simulation of such flows requires large computer memories and high computer speed to solve systems of coupled partial differential equations in several space dimensions and time. Computational methods for fluid dynamics have now matured to a point at which it is realistic to expect major breakthroughs in the 1980s and 1990s, provided that advanced computer resources are available. Some problems of modern fluid dynamics that can be effectively attacked, and possibly solved, with a new major national computational resource are

1. *Turbulence.* The highest-resolution three-dimensional turbulence code now uses 512 times 512 times 512 (or more than 100 million) modes to describe each velocity component; this simulation can barely be used to simulate three-dimensional inertial-range dynamics. These simulations are performed now at nearly the highest Reynolds number that can be obtained in a high-quality, low-turbulence wind tunnel. With more-advanced computational capability, a numerical fluid-dynamics laboratory is a realistic expectation.

2. *Transition to Turbulence.* The phenomenon of transition is essentially three dimensional, and so it requires great computational resources to solve. The problems of pipe flow, thermal convection, and

boundary-layer transition, for example, can be effectively solved and analyzed numerically. Numerical simulation offers the great advantage over laboratory experiment for these problems in that perturbations can be controlled accurately, so nonlinear effects can be isolated.

3. *High-Speed Flows.* Shocks are central to the dynamics of many physical systems. Examples include gas dynamics and implosion physics. Applications range from laser-fusion physics to sonic boom propagation to astrophysics. Many features of these flows, such as shock stability, interaction of shocks with vortical and acoustical disturbances, and multidimensional shock interactions are not well understood.

4. *Free-Surface Flows.* Applications include random surface waves, nonlinear interactions of surface and internal waves, Rayleigh-Taylor instability, and secondary recovery of oil. These problems involve the calculation of flows in highly complicated and convolved geometries.

One important aspect of fluid-dynamic calculations in the 1980s and beyond will be the presentation and analysis of multidimensional flow fields. Sophisticated techniques for the graphical presentation of results, including high-speed interactive graphics and videotape image output, will be required.

ASTROPHYSICS

Astrophysics problems are characterized by the extreme dynamical range of their physical situations. For example, in the observed jets of giant active galaxies, causally coupled hydrodynamic structure is seen over a range of 10^6 in linear scale, from parsecs to megaparsecs; and from time variations it is deduced that the actual scale of the machine powering the hydrodynamics is another factor of 10^3 smaller. The range of relevant densities in calculations of accretion onto a neutron star exceeds 15 orders of magnitude. The range of relevant times in stellar evolution calculations ranges from billions of years (for the lifetime of a star of medium mass) down to milliseconds (for the relevant dynamics of that star's supernova phase).

Astrophysics problems are also characterized by the range of different physical theory that they must include. Understanding the convective solar interior involves not just hydrodynamics but also magnetohydrodynamics, radiative transport, atomic physics (since the structure of ionization zones is of crucial importance), and probably also new understanding of the transition from order to chaotic behavior (which has recently become a unifying theme in many different areas of theoretical physics). The physics of supernova requires an intimate melding of shock hydrodynamics, nuclear physics, and (in some cases) general relativity.

One might list other subjects for study at the frontier of theoretical astrophysics that are particularly suited to a numerical approach: star formation, accretion-disk hydrodynamics, evolution of supernova remnants, galaxy formation (with the possible importance of such exotic species predicted by particle physics as neutrinos, and axions), spiral structure in galaxies, inhomogeneous baryon number creation in the early universe, formation of the solar system, common envelope binary stars and their evolution, and shock dynamics of the interstellar medium.

It should be evident that there is a great commonality of technique connecting almost all these problems: they generally require extensive numerical integration of the partial differential equations of hydrodynamics (almost always compressible hydrodynamics), or coupled radiative transport and hydrodynamics, in two—and sometimes three—space dimensions. The hydrodynamic computer codes that perform these integrations will be a key theoretical tool of the 1980s and beyond.

GRAVITATION THEORY

Only recently has it become possible to contemplate numerical integration of the full dynamical equations of general relativity. Already, the field of numerical relativity has answered important questions about the behavior of full relativistic systems such as black holes, neutron stars, and gravitational waves. The nonlinear, tensor nature of the Einstein equations in several dimensions makes obtaining relevant new analytic solutions virtually impossible. Hence, this is a field on which numerical solutions have a substantial effect. Important problems include the formation and collision of black holes, the likely central relativistic engine of quasars, and highly chaotic conditions in the early universe.

THE CHARACTER OF COMPUTATIONAL PHYSICS AS SCIENTIFIC RESEARCH

Computational physicists lead a triple life: first, and foremost, they must be fully conversant with conventional theoretical physics and the analytical techniques used by theorists; second, they must have many of the practical skills of an experimentalist; and third, they face the unique problems of computer programming.

Specialized physics knowledge must be combined with a general theoretical background in order to formulate problems to be studied on the computer, have the physical insight to guide the project through the myriad of decisions that must be made, and be sure that the results truly contribute to the solution of the problem being studied.

Equally important, tremendous analytical ability may be required to develop and understand the test cases and other methods used to establish the correctness of a computational procedure (both the algorithm used and the computer program that realizes the algorithm). Since the solution space of discretized equations is quite different from the solution space of partial differential equations, great care must be exercised in interpreting the solutions.

Furthermore, like experimentalists, computational physicists have an apparatus (the computer) whose recalcitrance must be overcome and whose idiosyncracies must be separated from those of the object being studied. Physicists must understand all sources of error in the calculations and design the code to keep all the errors from the approximations and numerical procedures under control. These include the problems of numerical instabilities, round-off errors, and, in the most extreme cases, the possibility of random hardware errors. They must deal with the practical limitations of limited funding, pushing the capabilities of the apparatus to its limits to permit carrying out computations at the forefront of physics research. Often considerable entrepreneurial skill is required to obtain the resources needed to carry out the research.

Finally, computer programming introduces problems. Many experimentalists also face this difficulty, but for the computational theorist the programming problems have led to special difficulties, including a great deal of misunderstanding and underestimation of the role and intellectual quality of computational physics. Computer programming and debugging is, in large part, a mind-dulling, menial task, in which hours and days and weeks are spent making trivial changes in response to trivial errors or figuring out how to format the output. Yet one must be able at any moment to apply the deepest analytical skills in order to understand an unexpected result or to track down a subtle bug. The computational physicist may worry a great deal that some unexpected combination of circumstances will cause the failure of a program in a real-life calculation despite the best efforts at debugging and testing the program. The computational physicist may need a detailed understanding of how the computer works, and of the nature of the compiler that was used for the program, in order to know how to test the program thoroughly and get it to run as fast as possible.

Underlying the practical problems of computer programming is the fundamental problem of readability of programs. For example, analytic theorists communicate the results of their work by writing papers. Other theorists can read these papers, rederive the results, and use both the results and the derivation in further work. In contrast, computer programs, the means by which computational results are derived, are by their nature unpublished, and complex programs can be read, if at all, at only substantial cost in time and effort. Furthermore, reading a large program is not enough; to achieve the understanding of another theorist's program that is comparable with the understanding of another theorist's analytic result requires building a totally new program to carry out the same computation, running both programs, and tracking down all discrepancies.

A basic need in any computation is the ability to increase its accuracy or expand its scope at the cost of increased computing time. Unfortunately, in virtually all the problems requiring large-scale computing there is a slow convergence problem: a substantial increase in computer time is required to achieve only a marginal increase in accuracy and scope. For example, one might want to decrease the grid spacing by a factor of 2 in a four-dimensional space-time calculation. This requires at least a factor-of-16 increase in computing time. To increase the accuracy of a statistical simulation by a factor of 2 requires at least a factor-of-4 increase in computing time. When the size of a matrix is increased by a factor of 2, the diagonalization time goes up by a factor of 8. These rapid increases often force computational physicists to search out the most powerful computing facilities available for their problems, even if this means going far from their home base and expending great amounts of effort on program development. The computational physicist learns to live with inadequate accuracy in his or her approximations and with an inability to explore many problems because the computing time needed is impractically long. Nevertheless, a great deal of excitement is felt at present among computational physicists: there are extraordinary developments taking place (or expected) in all parts of the computing world. If these are made accessible to physicists, they can open up new vistas for attack by computational methods despite the barrier of slow convergence.

POLICY IMPLICATIONS AND RECOMMENDATIONS

In May 1983, the National Science Foundation convened a workshop with scientists from diverse disciplines for the purpose of better defining the computational needs of the scientific community. Physics and its closely allied fields were significantly represented at this workshop. Most computational physicists would strongly support the six major findings of that workshop, which, in the following paraphrase, can serve as a useful first-cut set of guidelines for policy:

- Attitudes toward computers differ markedly between the older and younger generation of scientists. Older scientists may use—or have their graduate students use—computers for measurement and analysis tasks, but they consider the role and importance of computers to be limited. Younger scientists, who grew up with computers, view them quite differently as having, now or prospectively, a much broader and more essential role.

- There is a large gap between need and available support for minicomputers and microcomputers, attached processors, work stations, high-precision graphics, local area networks, and other local facilities required as part of the daily working environment of the research scientist.

- There is an immediate need to make supercomputers more available to researchers. The actual magnitude of the requirement is difficult to assess because lack of opportunity has artificially depressed the level of activity. The supercomputer initiative recently announced by the National Science Foundation is an appropriate and welcome response. It should substantially improve university involvement and leadership in research on computation and computers.

- Computer networks are necessary to link researchers, wherever they are, to large-scale computer resources and to one another. Establishment and support of such networks should be supported.

- The use of computers for research has passed a watershed. Computers are no longer simply tools for the measurement and analysis of data. Large computers, in particular, have become the means for making new theoretical discoveries. Academic research in computer architecture, computational mathematics, algorithms, and software for parallel computers should be encouraged as a means of increasing computational capability.

- Competition among computer manufacturers in education and research is resulting in favorable terms being offered to educational institutions for the acquisition of computers. Funding agencies have the opportunity to compound this effect on behalf of their large numbers of grant and contract recipients.

Scientists working in subfields of physics recognize their fields as being relatively further advanced in expanding the scope of their theory to include problems requiring large-scale computing. For example, the Panel on Plasma and Fluid Physics explicitly recommends

. . . a national computational program dedicated to basic plasma physics, space physics, and astrophysics, which will provide and maintain state-of-the-art technology appropriate to large-scale theoretical models and simulations. Such a program should ensure ready access to advanced computing on the basis of peer review.

In a similar vein, the Panel on Condensed-Matter Physics notes that

When properly carried out, simulations have the power to help us understand the physics of these difficult nonlinear problems. . . . Much of what we know about nonlinear dynamics comes from numerical calculations, and this will continue to be the case, especially as systems with several control parameters are investigated.

We think that it would be an error to assume that the need for enhanced computational capability lies exclusively, or even preferentially, in those subfields of physics whose rapporteurs, in their survey reports here, have chosen to highlight the role of computational physics. In our view, as exemplified by the examples in this chapter, the need for a vigorous and timely expansion of computational physics is universal, in all subfields of theoretical research. There is excellent work waiting to be done. There are discoveries waiting to be made by a new generation of theoretical physicists who have grown up with computers and for whom the computer is an extension of the mind, not just of the desk calculator.

7

The Interface Between Physics and Mathematics

INTRODUCTION

Interactions between physics and mathematics have stimulated discovery and invention throughout the history of both disciplines. Mathematics provides the language in which physics is expressed, and perceptions of orderliness in physical observations have stimulated the development of the mathematics with which to express these perceptions. Now profound developments at the mathematics-physics interface are stimulating remarkable activity in two areas that are reported here in essays by two leading participants in these dramas: These reports also reflect a new vigor that has developed at the physics-mathematics interface with the incorporation of computational physics, which potentiates theoretical analysis in many areas of mathematical physics.

This chapter supplements detailed accounts of advances in mathematical physics that appear elsewhere in this Physics Survey, particularly in the volumes on high-energy physics, on fluid dynamics and plasma physics, and on condensed-matter physics. Here some unique interdisciplinary developments are emphasized. The preceding chapter of this volume, on computational physics, amplifies another closely related interface.

The first development represents a revitalization of the historical relationship between understanding the physical laws of nature and the discovery of mathematical concepts. Newton's calculus was the mathematical vehicle needed to express his laws of motion; Riemann's non-Euclidean geometry and Hilbert's notion of a "function-space" provided the mathematical frameworks necessary for the expression, respectively, of Einstein's theory of general relativity and the quantum theory of Schrödinger, Heisenberg, and Dirac. It appears that we are just now entering a new phase of this process—one in which modern mathematical ideas, especially subtle geometrical and topological concepts, are playing a major role in the development of new theories of elementary particles and the structure of the universe, here reported as field theory and mathematics.

The second development is based on a new operational objective in analyzing the extremely complex behavior of physical systems in which nonlinearities lead to disorder and chaos. The new physical viewpoint focuses on understanding and representing the disorder instead of seeking only the simplest limits, but this objective demands a new mathematics. In fluid turbulence, biological pattern variability, population fluctuations, and the nonlinear dynamics of phase transformations it is the complexity of behavior that is the crux of the problems.

These remarkable developments depend on the expanding role of computers in physics and mathematics and, so far as we can see, have no historical precedent. We are not referring simply to what has been called computational physics, that is, the use of computers for data analysis and simulation of complex systems, for example, although those functions are certainly important parts of the picture. Rather, it seems to us that computers are beginning to make a profound change in the very way in which theoretical research is performed. We believe that computers, especially the flexible, interactive systems just now becoming available, are providing us with qualitatively more powerful ways of seeking answers to today's most challenging scientific questions. The emergence of what may in fact be a distinctively new research area at the interface between mathematics and physics bears implications for science policy and planning.

In the following sections of this chapter, we develop each of these two areas in more detail.

FIELD THEORY AND MATHEMATICS

Since the dawn of modern physics, developments in our fundamental understanding of nature have always involved the introduction of new mathematical concepts. The major revolutions in theoretical physics in the early part of the twentieth century were general relativity and quantum mechanics. These involved pre-existing mathematical theories (Riemannian geometry and Hilbert-space theory, respectively). The major theoretical advance of the late 1940s and early 1950s was the quantum theory of electricity and magnetism. This time the physicists were ahead of the mathematicians; the new advance in physics involved a complex and subtle mathematical structure, quantum field theory, whose role in pure mathematics is only now coming to be appreciated. In the 1970s, the big breakthrough was the understanding that weak interactions and nuclear forces can be successfully described by a class of theories known as non-Abelian gauge theories. These theories involve mathematical concepts whose role in pure mathematics has been grasped only in very recent times.

Because of the complexity of these theories, our understanding of them has developed only gradually, and many tools have proved useful. Perturbation theory, aided by the renormalization group, is the key to understanding the high-energy behavior of strong interactions; its success in this area is a central reason for our belief that quantum chromodynamics (QCD) is in fact the correct theory of strong interactions. Perturbation theory is even more useful in elucidating weakly coupled theories such as the standard Weinberg-Salam model of weak and electromagnetic interactions.

To understand low-energy aspects of strong interactions, unconventional tools have proved necessary. For example, large-scale computer calculations have furnished our strongest evidence that chiral symmetry breaking and quark confinement—two striking properties of the strong interactions as we observe them—are indeed predictions of QCD. This is an important novelty because large-scale computer calculations never before entered elementary-particle theory in such a fundamental way.

Our principal concern here focuses on the interrelation between physical theory and mathematical concepts. In this connection, another approach to gauge theories is of particular interest. This is the application to gauge theories of modern ideas in topology and geometry.

The first applications of topology to gauge theories were made in the early 1970s, in applications to weakly coupled, spontaneously broken gauge theories such as electroweak theories. It was found that even in these relatively simple theories, objectives exist that are outside the usual perturbative realm. The first such object whose existence in certain theories was uncovered was the string or vortex line (analogous to a phenomenon, the fluxoid, long known to occur in superconductors). The possibility that such objects may actually be relevant in cosmology has attracted growing interest, and experiments are under way that may confirm or refute this in the coming years.

Soon after the discovery of the string, an even more striking theoretical prediction was made—the magnetic monopole. Theories in which the strong, weak, and electromagnetic interactions are unified were found to predict the existence of magnetic monopoles as a result of simple topological considerations. This is now viewed as one of the key predictions of such theories. This prediction, too, has stimulated much experimental work.

These were the first applications of topology to modern gauge theories, but the subject really came into its own when applications of topology were made to strong interaction theory. In 1975, physicists Polyakov, Schwarz, and Tyupkin proposed the idea of the instanton— gauge field configurations in four-dimensional space-time with a property described by mathematicians as a nontrivial Pontryagin class. The instanton involved mathematical concepts far more subtle than had previously been relevant in physics, and it turned out to lead to two major advances in understanding strong interactions. It resolved an apparent conflict between QCD theory and experiment. This conflict involved a technical detail of meson masses, but resolving it was important because in 1975 it was the one clear conflict between QCD theory and experiment.

The other major role of instantons was to draw attention to an important but previously unrecognized problem, now called the strong CP problem (CP is the combined operation of mirror reflection and the transformation between particles and antiparticles). It is a basic fact of life that the physics of the strong interactions is left unchanged by transformations of isospin, strangeness, baryon number, charge conjugation, and CP. At first sight this is puzzling. Since none of the conservation laws just mentioned (except maybe baryon number) is a true symmetry of nature, why are these transformations conserved by the strong force? One of the triumphs of QCD was to explain this; QCD is so constructed that it is impossible for QCD forces to violate the conservation laws mentioned. That is how matters seemed to stand before the era of the instanton. With instantons, it became clear that although isospin, strangeness, baryon number, and charge conjugation are indeed automatically conserved in QCD, CP is not. Why the strong interactions conserve CP thus emerged as a major mystery. It soon became clear that CP conservation by strong interactions can be naturally explained if a new light particle, the axion, exists. The original form of the axion theory has been ruled out by experiment, but a variant, the weakly coupled, or invisible, axion, is currently the subject of active experimental search.

By this point, modern mathematical ideas had entered relativistic quantum field theory to stay. The subsequent development has been complex, with new applications of mathematics appearing in many areas.

Looking to the future, it is clear that ideas from modern mathematics will continue to play a role in uncovering the secrets of gauge theories. It also seems likely that quantum field theory may come to play an influential role in mathematics. Still to be explored are fascinating theories—supergravity and supersymmetric string theory—whose complex mathematical structure is still not understood. It will not be surprising if ideas from modern mathematics play a role in shaping the next generation of fundamental theories.

CHAOS AT THE INTERFACE OF MATHEMATICS AND PHYSICS

Mathematics is the language by which physics is expressed. Traditionally, physical insight isolates the mechanisms that account for a phenomenon, and then a talent of a perhaps more abstract nature transcribes it into binding mathematics.

The methodology of basing mathematics consistently on the perceived workings of the world originated in the school of Pythagoras, for whom the perceived regularities or harmonies of the physical world became comprehensible through the invention of the mathematics that expressed those notions. Similarly, modern physics grew to its present form with Netwon's invention of the calculus as the intuitive mathematical vehicle needed to express Galileo's notions of inertia. During such periods of paradigmatic construction, the boundary between mathematics and physics has been diffuse or nonexistent.

At present, in consequence of the availability of interactive digital computation, another such period in which an extension of the reach of analytical methodology is sought is perceptibly beginning.

Theoretical physics is a discipline in the highest analytical tradition, striving to identify a small number of the most elementary principles from which a large body of phenomena can be reconstructed. Toward this end only the most symmetric and elementary solutions are investigated to determine whether the assumed underlying principles are substantially correct. The technique of finding such simple solutions is an artistic skill that a successful theorist acquires. In contradistinction, the availability of more general complex and asymmetric solutions is beyond the range of existing mathematical methodology. A time-honored example is the problem of fluid turbulence, in which there is little doubt that the fluid field equations, expressing elementary Newtonian transport, are substantially correct, while the ability to comprehend these equations' predictions for turbulent flow is almost totally lacking.

Historically, there is little to suggest that highly complex solutions will emerge once the principles are all recognized—that is, a theory can be true, yet humanly irrelevant. The oustanding example of this notion, in the context of physics, is the theory of gases. At the most fundamental level, Newton's equations for a very large number of particles could yield the kinematics of every particle, yet pursuing the problem of the behavior of gases this way is of almost no interest. Rather, a new theoretical discipline, that of statistical physics, had to be created to make this system comprehensible. The intuitions that underlie this discipline only marginally make contact with the Newtonian picture; instead they rely more significantly on the separate insights of thermodynamics.

In just such a vein, it appears most likely that the significant comprehension of fluid turbulence and other complex phenomena will be achieved only after a more suitable theoretical discipline emerges that is aimed in a direction fundamentally different from the simple solutions of the fluid equations. In this pursuit, a parallel to the fine intuitions of thermodynamics is regrettably not yet available, so that the intuitions, the methodologies, and the appropriate mathematical framework must be simultaneously erected. That an attempt should appear to be at all feasible now, when attempts over the past 100 years have largely failed, is a consequence of modern computation.

Beyond the physics of fluids, the general problem to be addressed is the analytical description of all those systems that require a seemingly astronomical amount of information in order to specify their states. Thus, the fluid equations require as initial data a velocity field specified at each point of space in order to see beyond as the system advances in time. A similar requirement exists for the evolution of a distribution of matter under the gravitational-field equations. Biological systems, with their myriad details, quite possibly exceed the reach of any hard analytical science, while the most preliminary studies of neural nets pose problems of immense initial data evolving under a system all of whose parts are mutually interacting. For solutions of these problems, a local Newtonian paradigm seems fundamentally misaimed. Common to them all is the need to determine variables other than the usual local kinematic ones that can serve as a framework for theories to allow human intuition to come to the fore. Such a pursuit demands new mathematics and deprives the notion of a surgical separation of physics and mathematics of meaning.

During the past decade, the principal method of dealing with complex objects has become the scaling idea of the renormalization group. This is simply to say that a highly complex structure can result from the recurrent action at all length scales of a simple, largely scale-invariant process. By way of analogy, a fern that presents a boundary, the equation of which is essentially impossible to write down, achieves its form by four or five levels of self-similar growth. Indeed, the theory of its form is the genetic encoding of an elementary growth process: its final form contains an immense number of bits of local information, whereas relatively few instructions are required in order to specify its growth process. Perhaps identically, dissipative systems such as fluids relax to configurations that, while locally requiring immense amount of information to describe, can be expressed through small amounts of global scaling information. For an approach to be feasible, the kinematics of such a system should embrace these scalings rather than focusing primarily on local motions. Although no such theory is as yet extant for fully developed turbulent flow, there are experimental indications that such a description may be feasible. The onset of turbulence for some flows has been understood in precisely and only this way. A new body of mathematics precisely aimed at representing this physical intuition has been erected; it differs radically in content from the detailed local fluid field theory; yet it exactly determines consequences of the field equations that are now just becoming available as empirical output of simulations.

Thus, a new set of problems is becoming ripe for investigation, and the effort in elucidating them rests precisely at the dividing line between two disciplines. But mathematics and physics do not easily merge at those points where neither alone is providing sufficient support. In order for a mathematical formalism to cross react with the intuition of physics, both must be grounded in quite specific frameworks. Therefore, it is worth considering how the process is to be facilitated.

Traditionally, a physicist can discern the necessary tools and then consult with the mathematicians to discover how to proceed. Here this is not possible. First, the discerned phenomena and tools are already skewed toward mathematics, and second, the mathematicians do not possess the requisite knowledge. Thus, something much closer to ongoing collaboration is necessary. To this end, the growing number of centers for nonlinear studies provide a meeting ground for daily discussion and collaborative effort. The growing efforts to redesign classical mechanics with a concomitant exposure to more modern mathematics introduces students to the ubiquitous fact of nonlinearities. Similarly, dynamical systems courses and seminars have been exposing mathematicians to more of the arena of physical problems that are demanding new mathematics. Support and proliferation of these centers and courses seem at present to be the best way to proceed.

The actual methodology of investigating nonlinear systems exhibiting complex excitations has its key in interactive computation. One is no longer surprised at seeing a serious mathematician in front of a graphics terminal. Five years ago, a mathematician could risk losing contract support if funding was requested to purchase such a terminal. Simply, it is now accepted that intelligent exploratory computation makes its users more intelligent than they could have guessed they would be. A great mathematician will regularly say that once one knows what is true, the proof is easy.

This goes against the misconception that the scientist poses hypotheses and then, armed with the scientific method, disproves or adds credence to their validity. Of course, the essence of research is identifying significant hypotheses—whether in physics or mathematics. And this is the essence of interactive computation. Once a setting is uncovered that possesses hints of orderliness in a context of complexity, one can manipulate the object until seeds of intuition appear. Our inheritance of experience with simple systems is strikingly empty of images, intuitions, and methods for dealing with nonlinear problems of complexity. We know almost nothing of the workings and accustomed regularities of such systems. And to proceed we must come to know them intimately.

But how does one submit a batch job to a computer facility to explore what one does not know? This is not intended as a conundrum. Rather, it is simply impossible. The fact is that modern computers are fundamentally the tools of banks and businesses. For scientific purposes, they serve well for the detailed runs of solutions to elementary models for which the accurate results of an understood qualitative behavior are sought. Time-sharing terminals basically are a facilitation of this batch-oriented work. In order to illuminate new settings, in which the interesting results are unexpected surprises encountered after educated trial and error, something much more like an analog computer with knobs and pictures is required. Few such computers are generally available, and when they are available, they are not generally appreciated. One needs effectively stand-alone systems of perhaps only modest speed that can operate close to interpreters, allowing code to be written as an executing program's output inspires new thoughts; that allow algorithms to be modified on the fly; and that demand the least possible in the way of a fully preconceived program. Five years ago, funding for such a facility (at a budgeted cost of about $20,000) would have been difficult to obtain. Fortunately, this situation has changed appreciably since then.

A UNIX operating system by itself is insufficient for dealing with such problems. Rather, we recommend that a high-level interpreterlike operating system or subsystem be made easily available. This system should be capable of supporting machine interruptions that can be obtained through simple software commands. It is this addition that allows a digital computer to emulate an analog device. Such a machine, running 1000 times slower than a major mainframe, may be orders of magnitude more powerful than the latter for purposes of solving nonlinear problems.

In summary, we note that there now exists a growing subject matter that defies the traditional division of physics and mathematics and adds to them a requirement for innovative computation. Its successful pursuit almost surely requires close working arrangements among previously disparate practitioners. Indeed, each must come to be some of the other in the interdisciplinary style characteristic of all the interfaces recounted in this volume; this process should start no later than graduate school. It requires as its basic tool flexible, interactive computing facilities that must be more readily available.

8

Microelectronics and Physics

INTRODUCTION

The modern world is in the midst of a second industrial revolution, a new information age superseding the old industrial age. As Figure 8.1 shows, today nearly 50 percent of the U.S. work force is engaged in the gathering, processing, and dissemination of information. The rise to predominance of the information segment of our economy has occurred during the past 30 years. It has been directly fueled by the birth and exponential growth of modern electronics during the same period. Today electronics is a more than $100 billion industry in the United States. But modern electronics is more than a healthy segment of our gross national product, more than the technological basis of the information component of our economy. The new microprocessor era of electronics is beginning to exert immense influence on the nature and direction of all segments of our economy and our society as a whole.

If microelectronics is the obvious linchpin of this new age, the obvious foundation is the past half-century of modern physics research. The bulk of this physics foundation comes of course from condensed-matter science, particularly the subfield of semiconductor physics, but it includes fundamental contributions from atomic-, nuclear-, and plasma-physics research as well. Although microelectronics today appears to be a scientifically mature technology, it remains closely based in physics (and chemistry), a technology whose current status and future directions have

134

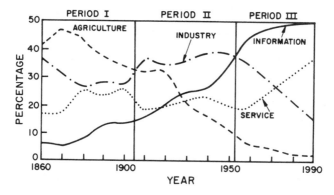

FIGURE 8.1   Changing composition of U.S. work force. [After A. L. Robinson, "Electronics and employment: displacement effects," in T. Forester, ed., *The Microelectronics Revolution* (MIT Press, Cambridge, Mass., 1981), p. 381.]

been and will be profoundly affected by the scientific research efforts of the recent past and the immediate future.

Thirty-five years ago physics research provided the scientific basis for the birth of microelectronics. Today condensed-matter physics and materials research are continually refilling the well of science from which microelectronics draws for the continuing evolution and development of the technology. At the same time, the general potential for technological relevance in combination with intrinsic scientific merit continue to give particular vitality to those areas of physics that have the potential of supporting microelectronics technology.

This chapter examines the interactions of physics research with microelectronics technology. It will not extensively survey the current status of the technology and will not review the central role that electrical engineering has played in bringing this technology to its current high level of success. Rather the focus here is on developing a better understanding of the complex and sometimes subtle ways in which physics research can still profoundly affect the development of a well-established technology. Thus, this chapter acts to complement that devoted to optical information technologies (Chapter 9), which are just now emerging from the research environment.

FROM TRANSISTOR TO ULTRALARGE-SCALE INTEGRATION

Modern electronics and semiconductor physics both began with the mid-twentieth-century invention of the transistor. In terms of organization and approach, this invention is a premier example of a directed physics research effort, and in terms of long-term impact this invention is, arguably, the best example of unforeseen consequences of physics research. The objective of this effort was to develop a low-power, more-reliable replacement for the vacuum tube. But the transistor was soon followed by the solid-state computer and, a little more than a decade later, by the integrated circuit. These key events have led to the electronics revolution of today. This revolutionary change and the seminal role that the transistor would play in bringing it about could not have been imagined by its inventors.

The changes that the transistor wrought in physics have been almost as revolutionary. The transistor, founded on fundamental understandings of solid-state physics, promoted an explosion of physics research activity that has continuously been intertwined with technology, at times contributing to technology advance, at times drawing from technology to advance physics research. Whereas early transistor technology was implemented with germanium, scientists and technologists explored and created a variety of other semiconductors, purified them, processed them, and studied their fundamental properties. They eventually found that silicon, with its higher band gap and excellent oxide thermal insulator, was the ideal basis for modern integrated circuits.

Along the route to that discovery, many subfields of semiconductor physics and specialized technologies developed that survive today, thereby providing a base for extensions and alternatives to current applications. The science underlying emerging optical information technologies also evolved in this manner. On the other hand, such recent fundamental discoveries as the integer and fractional quantum Hall effects are a direct outcome of the technology and science of microelectronics. Today the intimate connection between new physics and new technology remains at least as strong and clear as it was 35 years ago.

Decade of Very-Large-Scale Integration

Physics research provided the fundamental foundation of modern microelectronics technology, and it has been an essential factor in the rapid and continued growth of this technology. The major development in microelectronics during the past decade has been very-large-scale integration (VLSI). Since the invention of the integrated circuit (IC) in 1962, the complexity of IC chips has grown at essentially an exponential rate. The latter 1970s initiated the era of VLSI in which this complexity has reached a level of more than 100,000 components per IC chip. Today, memory chips of less than 1 cm^2 in size and capable of storing a quarter-million bits of information (256K RAM) are rapidly becoming a commodity product. Microprocessor chips with nearly a half-million transistors are being installed in relatively low-cost professional computers.

These remarkable developments have been accomplished in part by increasing the size of the chip. But the biggest factor has been the exponential decrease in the minimum feature size (linewidth) of the devices on the chip. Since the 1960s minimum linewidths have been decreasing at an average rate of a factor of 2 every 2 to 2 1/2 years. Currently, standard ICs are being produced with linewidths at the 2.5-μm-scale. Now we are beginning to see the introduction of VLSI chips with linewidths of 1.25 μm and component counts in excess of 10^6. The 1-μm linewidth/10^6 component circuit is beginning to be referred to as ultralarge-scale integration. This continual scaling down of device dimensions yields continual improvement in performance—smaller devices, properly scaled, are faster devices—while increased chip complexity yields both greater system performance and lower cost per system function. The result has been even more powerful electronic circuits available at much lower cost, i.e., the microprocessor era. If the historical trend continues, 1990 will see ICs with millions of components per chip and with linewidths approaching 0.5 μm.

VLSI devices and circuits continue to operate on the principles of electron-device physics that were worked out by semiconductor physicists in the 1950s and early 1960s and that were employed in the first ICs. Current indications are that this miniaturization will continue until device dimensions of the order of 0.25 μm are reached in the next decade. This is based in large part on the computation of the fundamental physics limits to the continuation of progress with existing materials and processes. Thus the major contribution of physics to VLSI *during the past decade* has not been to provide new electron-device concepts but rather to provide the basis of the technology developed to achieve this scaling and manufacturing capability.

The rapid advances in VLSI have been and will continue to be critically dependent on the development of new materials, on new materials-analysis techniques, on the development of new and more controllable means with which to process materials, and on the continued evolution of higher-resolution methods of patterning devices and circuits.

The intimate connections between processing technologies and research are readily apparent. For example, over the past decade, as device dimensions have rapidly shrunk, the scaling requirement for proportionally shallower doping layers and sharper boundaries between doped and undoped regions in semiconductor structures has resulted in the transformation of ion implantation from a tool for scientific investigation to the preferred industrial technique for doping semiconductors. Since ion implantation acts to damage semiconductor crystals, introducing undesirable defects, new and powerful analytical techniques such as Rutherford backscattering spectroscopy (RBS) have been successfully employed to characterize these effects.

The necessary removal of ion-implantation damage without smearing out doping profiles necessitated the development of laser annealing. When lasers became widely available, they were quickly and effectively employed as spectrographic probes by semiconductor physics. But laser beams, when intense enough, can also melt and recrystallize semiconductor surfaces. In the course of the study of such interactions many desirable effects were observed, such as crystallinity, sharp definition of dopant profiles, and controlled intermixing of layers. Some of these effects may occur within the solid state, i.e., without melting of the exposed surfaces. Although these and related effects were discovered and in part analyzed with lasers, recently strong incoherent sources (bright lamps) were developed to reproduce many of the useful laser effects in ways more compatible with manufacturing techniques. Commercial prototypes of lamp annealers for semiconductor manufacturing lines are now becoming available, and rapid thermal annealing is expected to play an important role in microelectronics fabrication over the next decade.

The remarkable decrease in circuit size illustrated in Figure 8.2 has also led to the necessity of developing new materials for microelectronics. Polycrystalline silicon (polysilicon) is rapidly being replaced as the gate electrode structure in metal-oxide-semiconductor field-effect-transistor (MOS-FET) devices by polysilicon/metal-silicide thin-film structures. This replacement is mandated by the requirement of higher conductivity for the smaller electrode structure. More-stable refractory metals are likewise replacing aluminum as a thin-film interconnection material. Again the development of these new materials has depended on processing techniques that originated in basic research. Particularly in the case of the silicide materials, various microanalytical techniques, especially RBS, have also been crucial in characterizing these new thin-film materials and establishing effective fabrication processes.

VLSI requirements have increased the importance of advanced microanalytical techniques to the electronics industry. In addition to RBS, this past decade has seen such techniques as Auger electron

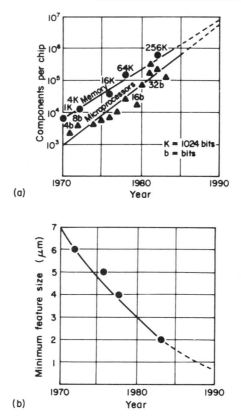

FIGURE 8.2   (a) Increase in density of ICs over the past decade. (b) Decrease in the minimum IC feature size over the past decade. [After W. L. Morgan, *Semiconductor International* (Cahners, Des Plaines, Ill., May 1984), p. 101.]

spectroscopy and x-ray photoemission spectroscopy become auto-mated (with the use of microprocessors) and move from the surface-science research laboratories to the microelectronics development and production lines. These amazingly sophisticated instruments are pro-viding information vital to the development of new electronics mate-rials and to the maintenance of process control in microelectronics fabrication. High-resolution transmission electron microscopy (TEM) is likewise becoming an increasingly more important analytical tool for the study of interface atomic structure and of other microstructure with nearly atomic resolution. Secondary-ion mass spectroscopy is also finding widespread application in the electronics industry.

Indeed, one of the most striking developments of the past decade has been the ubiquitous infiltration of scientific instrumentation into the high-technology workplace. At the same time, the successful commercialization of these various microanalytical techniques has made this instrumentation widely available to the scientific community to carry out its basic research mission.

Continued evolution of VLSI requires continued progress in replicating ever smaller, ever more complex patterns in an accurate and economical manner. Optical lithography is expected to continue for some time to remain to be the dominant means by which microcircuits are patterned. But in the past decade electron-beam lithography (EBL) has become the technique of choice for generating the master pattern (mask), which is then optically replicated thousands of times over.

Although current production economics do not favor EBL for direct production of VLSI circuits (direct write), this technique is ideally suited for the research and development and special-purpose production environment. This application of EBL began with the invention of the TEM by physicists, was followed by the development of the scanning electron microscope by electrical engineers and applied physicists, and then was brought to its present highly developed state by the combined efforts of many scientists and technologists. As circuit dimensions continue to shrink, the importance of EBL will continue to grow. EBL was recently demonstrated to be capable of creating structures as small as 4 nm. Clearly the limits of microlithography are quite fine indeed.

As commercial VLSI circuit design rules approach the 1-μm barrier, the limits of optical lithography begin to be felt. (These limits may be extendable in several ways, including the use of the recently developed excimer laser.) Consequently, soft-x-ray lithography is beginning to be more seriously considered as the submicrometer pattern-replication technology. X-ray lithography has in fact been demonstrated to be capable of producing sub-10-nm structures. The technical problem is now to make x-ray lithography commercially feasible. A major concern is to develop a source of soft x rays intense enough to minimize the time required for pattern exposure at submicrometer linewidths; low-energy (1-GeV) electron storage rings and pulsed plasma sources may provide the answer. Thus technology initially developed for high-energy physics research and then adapted to provide synchrotron radiation for fundamental research may in the near future play an important role in microelectronics.

Focused-ion-beam systems are currently under intense development by scientists and engineers for future application as a submicrometer pattern-generation tool. In principle, focused ion beams offer significant advantages over electron beams for lithography, for localized submicrometer etching, and for local ion implantation. Although major difficulties remain in the areas of ion-source development and ion optics, focused ion beams appear likely to be a major area of technical development in the next decade.

Just as the rapid evolution of VLSI has created the demand for major advances in microlithography, it has also demanded major changes in the way in which microlithographic patterns are transferred (etched) into electronic circuits. In the early 1970s the standard pattern-transfer technology was wet chemical etching. Today the demand for micrometer lines with submicrometer acuity has resulted in the rapid development of reactive ion etching (RIE) as the key pattern-transfer process. RIE is a low-pressure plasma technique in which reactive ions and radicals erode the exposed surface of a substrate in an often complex physical-chemical process. The etching can be highly anisotropic, and it has been demonstrated that under the proper conditions RIE can etch structures as small as 100 Å. Developing successful RIE processes and establishing a basic understanding of the fundamentals of RIE mechanisms has been and continues to be a major activity in the field of microfabrication. Surface physicists and surface chemists have made fundamental contributions in this area, and the rapid development of the technology has been critically dependent on the use of a variety of microanalytical and spectrographic tools.

Symbiosis

Just as recent developments in microfabrication technology have been based on physics research, certain areas of physics, particularly condensed-matter physics, are now utilizing this new microfabrication technology to carry out new fundamental research. The ability to produce structures on a nanometer scale has facilitated recent investigations into such areas as conduction electron localization, nonequilibrium superconductivity, and ballistic electron motions. Microscience is becoming an area of increasing activity in solid-state research laboratories. Indeed, one of the major reasons for which the National Science Foundation established a National Submicron Facility in the late 1970s was to help to make this impressive microfabrication technology available to scientists for fundamental research.

The interactions between microfabrication technology and microscience are of course quite complex. In fact, it is often the scientist who is now pushing the microfabrication technology to finer limits. It remains for the technologists to make use of these limits practical—a much more difficult task.

OTHER TECHNOLOGIES

Silicon devices and integrated circuits will, for the foreseeable future, dominate the electronics industry. The dynamic, rapidly evolving, highly successful nature of silicon technology gives no indication that it is about to be supplanted by another. On the other hand, continual interactions between physicists and technologists have engendered alternatives to conventional silicon technology. These other materials may successfully compete with silicon in the area of high performance or fill special applications niches—niches that may be a small part of the current electronics industry but that may be of large economic and societal importance.

Gallium Arsenide

At the beginning of the semiconductor era, gallium arsenide (GaAs), which does not not occur in nature, was synthesized along with many other semiconductor compounds with the hope, now fulfilled, of creating special properties not realizable in silicon or germanium. For the past two decades GaAs has been of technological importance because of its high electron mobility and its optoelectronic properties, the latter being a result of the direct (momentum-conserving) band gap of GaAs. The direct-band-gap concept and the need for it in optical emission devices such as diode injection lasers (a key to optical communications technology) came from semiconductor physicists.

The high mobility, or equivalently the high charge-carrier saturation velocity, of GaAs early on made it the semiconductor material of choice for high-frequency, microwave applications. Since then the continual development of very-low-noise GaAs FETs, high-speed analog-to-digital (A-to-D) converters, and other GaAs devices has had a fundamental impact on the development of commercial and military communication systems (and on radio astronomy). Satellite communications are particularly dependent on high-performance GaAs electronics. Direct satellite-to-home broadcasting is rapidly becoming technologically and commercially viable.

Following the early successful synthesis of GaAs, the development of new growth and processing methods, particularly epitaxial techniques, opened promise of new device structures. In recent years this promise has been amply fulfilled but, as usual, in often unexpected ways.

Molecular-beam epitaxy (MBE) was developed to grow exquisitely controlled thin films of high-quality compound semiconductor materials. A major focus of the initial MBE effort was to produce alternating layers of two different semiconductors, GaAs and GaAlAs, each only a few atoms thick. The resulting superlattice and quantum-well structures manifest new fundamental physical phenomena and permit speculative applications in devices such as high-frequency oscillators, photodetectors or photovoltaic converters, special function lasers, and electron multipliers.

But the GaAs superlattice research effort also strongly accelerated MBE technology. MBE-type techniques in the past few years have begun to see wide application in thin-film research. A cottage industry has sprung up, producing nearly atomically thin, alternating layers of materials for fundamental studies in superconductivity, electron localization, and magnetism and for such applications as mirrors in x-ray optics. At the same time, MBE has joined metallo-organic chemical-vapor deposition as the technique of choice for producing high-performance compound semiconductor systems for microelectronic and optoelectronic application. Experimental programs in silicon MBE are also flourishing.

One unexpected result of the GaAs superlattice effort was the discovery of modulation doping. By growing alternating, ordered heteroepitaxial layers of GaAs and GaAlAs, it was found that the donor atoms that provide the electrons in the electrically active GaAs layer could be placed in the adjacent GaAlAs layer. Thus, unlike in a conventional semiconductor, the donor ions could be removed from the conductive channel where they would otherwise act as electron scatterers and thereby limit device speed. The exceptionally high electronic mobility of modulation-doped GaAs-GaAlAs heteroepitaxial layers has recently been used to produce FETS with (unloaded) switching times of the order of 10 ps, as shown in Figure 8.3.

At the same time, modulation-doped GaAs layers are being used for the study of basic physics phenomena. Of particular note are the quantized Hall effect (recognized by the 1985 Nobel prize) and the fractional quantized Hall effect—two recently discovered, related phenomena that give precision measurements of fundamental physical constants and a potential new electron-liquidlike state of matter, respectively. Sometime in the future these new phenomena may well find application in electronics technology.

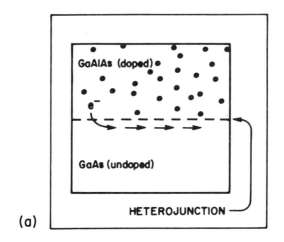

(a)

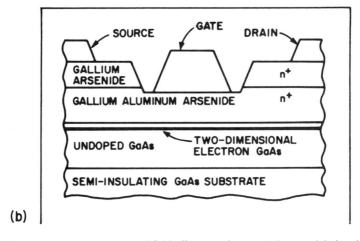

(b)

FIGURE 8.3   Heterostructures and field-effect transistors. (a) In a modulation-doped heterostructure, electrons move out of the doped GaAlAs layer into the undoped GaAs layer where their potential energy is lower. The ionized donors left behind attract the electron back toward the AlGaAs layer. The result is the formation of a narrow, two-dimensional, free-electron gas layer in the GaAs layer close to the interface. These electrons can have an extremely high mobility or equivalently an extremely long mean free path between collisions. (b) A schematic representation of a modulated doped field-effect transistor. These devices are also called high-electron-mobility transistors or two-dimensional electron gas transistors.

GaAs is a much more difficult material to work with than silicon. Wafers are much harder to prepare, the material is more susceptible to damage during processing, and GaAs is not endowed with a rugged, high-electrical-quality native oxide as is SiO_2. Consequently, GaAs has not traditionally been seen as an effective competitor to silicon in integrated circuits. Only recently have the physics phenomena been shown to be so promising as well as easy to implement in IC configurations that the necessary technology efforts have been directed to GaAs-wafer preparation and processing. The result is a worldwide race to develop both conventional and modulation-doped GaAs integrated circuits. This in turn has resulted in a worldwide effort by those who favor silicon ICs to meet this new challenge.

Other, more exotic, semiconductor materials may be even more difficult to prepare and use than GaAs. Other III-V compounds, such as GaP and InSb, either alone or alloyed, exhibit properties that make them attractive for special applications, e.g., for lasers or photodetectors matched to optical fiber characteristics. Similar statements can be made about II-VI compounds and alloys, of which HgCdTe is a notable example because its narrow band gap makes it suitable for use in far-infrared detectors. There are fundamental physics as well as applications-oriented activities on these and other exotic semiconductors. Because of the formidable fabrication problems, research in these areas is not close to providing mass microelectronics applications. But the lesson to be learned about GaAs is that when the phenomena that push beyond the limits of current technology (as GaAs does with respect to speed and power consumption) can be demonstrated in an integrated circuit, then technology resources can be successfully brought to bear on the fabrication issues.

Josephson Junctions—A Superconducting Computer?

The Josephson effects and other manifestations of the macroscopic quantum nature of superconductivity are among the most beautiful and novel effects in physics. Pursuits of their technological application have continued to be most exciting activities in the past decade. Central to rapid progress in this area has been the highly successful development of microfabrication technology for micrometer- and submicrometer-sized Josephson junction structures of high quality, uniformity, and reliability. Magnetic-sensor and millimeter-wave detectors have been developed to sensitivity levels nearly at the quantum limit set by the uncertainty principle. We are now seeing growing application in such diverse fields as radio astronomy, biomedical research, and magnetic monopole surveys. Josephson microstructures are also being used to address a variety of fundamental questions in condensed-matter physics, including chaotic behavior in physical systems and two-dimensional phase transitions.

The area where Josephson technology could have by far the most impact is in digital applications. Over the past decade Josephson digital technology has received considerable attention. Single-chip, 10,000-junction, 1-kbit RAM circuits have been produced.

Very-high-speed logic circuits and a packaging approach have been successfully demonstrated in another effort. Fully loaded logic delays as small as 10 ps have been demonstrated. Even higher speeds are possible. In addition to the high speed and low power dissipation that Josephson technology offers is the important benefit of lossless superconducting transmission lines for interconnections. Research and development efforts on Josephson junctions have demonstrated what can be accomplished and have shown that this technology is perhaps the ultimate in terms of potential performance.

Recently Josephson electronics has had strong competition, at least for the short term, from the major advances in high-speed, modulation-doped, high-electron-mobility transistor (HEMT), GaAs circuits. Moreover, the realization of the potential of Josephson technology has suffered a setback recently because of difficulties encountered in manufacturing a complete large-scale high-speed logic and memory package for application as a high-performance general-purpose computer. The complete Josephson-computer package approach was considered necessary since all subsystems of such a computer must be in a 4-K environment for overall system performance to be maintained.

Failure of this all-or-nothing approach dictates, for the near future, that Josephson-based electronic circuits will most likely be limited to filling niches where the highest speed and performance are required. Examples of such niches are analog signal convolvers, A-to-D converters, transient signal sampling circuits, and special-purpose signal processors. The eventual movement of superconductor electronics out of these special niches into the larger arena of high-throughput computers will depend on a combination of scientific and technological developments. One such development would be a significant improvement in the manufacturability of Josephson integrated circuits, which means gaining better understanding and control of the crucial tunnel-barrier formation process. Another positive advance would be the invention of improved superconductor digital circuits; particularly desirable would be the development of a millivolt-level cryogenic device with transistorlike properties. The adaptation of semiconductor devices to interface with superconductor devices and dissipationless transmission lines in a 4-K environment would be another important development. There are numerous opportunities for research in these areas. When the current high rate of improvement in conventional semiconductor circuits eventually declines, such research will be essential to further advancement of high-speed, supercomputer circuits.

Amorphous Semiconductors

Amorphous semiconductor physics, originally motivated by scientists' desire to understand the effects of extreme disorder in semiconductors, received a big stimulus in the late 1960s when switching devices were proposed that utilized chalcogenide glasses. Chalcogenides, compounds or alloys that contain sulfur, selenium, or tellurium, had also been the materials of choice for the early physics studies because of their relative ease of formation in the glassy or amorphous state. Later efforts with germanium and silicon, and the search for fundamental crystalline analogies within the amorphous state, led to the successful *p*- and *n*-type doping of amorphous silicon and subsequent demonstrations of *p-n* junction diodes, photovoltaic cells, sensitive photoconductors for vidicons and electrophotography, thin-film transistors, and other devices. Here is another example of the synergy between science and technology. The interesting twist was that despite the early technology impetus of the chalcogenides, when scientists turned to silicon and germanium to find simpler and more fundamental phenomena and properties, they found a rich field of applications. Amorphous silicon is now the basis of a host of commercial devices in energy conversion and imaging. More applications appear on the horizon. When the scientific breakthrough in amorphous silicon came, many technology laboratories actively using chalcogenides were able to redirect their efforts quickly by virtue of their knowledge of and conceptual connections to the physics of amorphous semiconductors. These connections are still active today.

SCALING DOWN—LIMITS TO MINIATURIZATION

In order to continue the improvement in the performance of integrated circuits, to continue the rapid decrease in the complexity/cost ratio of ICs, and to increase the performance and extend the cutoff frequency of microwave transistors, the technological imperative is to

FIGURE 8.4   A bright-field electron micrograph of an array of holes directly etched into an 80-nm-thick AlF_3 film by a subnanometer-diameter beam of 100-keV electrons. The hole diameters are between 1.5 and 2.0 nm with average center-to-center spacing of 4.0 nm. The walls of the holes are plated with aluminum. (Micrograph courtesy of M. Isaacson, Cornell University.)

continue scaling down the size of electronic devices, well into the submicrometer regime. The rapidly developing microfabrication technologies are clearly providing the technical means of fabricating smaller (and more complex) devices and circuits, as seen in Figure 8.4. But as electronic structures move well into the submicrometer regime, new demands are being placed on the scientific knowledge that forms the basis of microelectronics. We shall briefly examine several topics of critical importance to present-day and future microelectronics.

Materials

As device dimensions continue to shrink, thinner, more controlled, higher-quality materials layers must be produced, and more-complex structures will be required. Recently developed materials growth techniques, such as MBE, low-pressure chemical vapor deposition, cluster beam deposition, and graphoepitaxy will require further refinement and improvement. More demand will be seen for materials modification techniques such as ion implantation and rapid annealing.

Completely new growth and modification techniques are likely to be necessary to meet future needs.

As electronic structures become more and more miniaturized, current understanding of the concepts of impurity diffusion/migration, microstructure, and surface properties in these materials becomes insufficient. Although defects such as microinclusions or precipitate structure may be negligible in a large structure, the physical dimensions involved can be expected to be a large fraction of the size of an ultrasmall device structure, and such defects will necessarily become dominant.

The nucleation of a thin film is governed by the detailed non-equilibrium (in general) thermodynamics of the system and includes critical questions about the kinetics of the growth process, the role of weak and/or metastable atomic bonding, molecular motion forces, and cluster formation during growth. These detailed processes are not well understood. Thus, in the range of dimensions expected for ultrasmall devices, current models for such effects as defect formation, chemisorption, segregation/agglomeration, atomic stability, solid-phase reactions, and microinclusions are likely to be inadequate. Effects that may be unimportant today may well become dominant on the ultrasmall scale.

The growth of epilayers on the less than 0.1-μm thickness scale is a problem owing to the tendency for nonuniform growth. Moreover, the inclusion of some impurities tends to destroy the structure and upsets the stability of multilayer structures. In multicomponent films, inadequate knowledge of kinetics leads to tendencies toward the nucleation of microstructure or ordered clustering within the film or dendritic growth. The presence of defect clusters, both at the surface or interface and within the film, becomes a problem not only because of the size of the defect cluster but also because of the presence of localized excitations that may dominate the electronic properties. In the case of ion implantation, the annealing and regrowth of the implanted layer are not understood well enough to characterize microinclusions or residual damage on the size scale of interest. Enhanced diffusion along grain boundaries or dislocations produces nonuniformities that can be detrimental to performance of a small device.

The serious problems in epilayers or in implanted layers are compounded by the problems of the substrates themselves. In silicon, the basic material, problems still exist in resistivity nonuniformity, defects, and structural properties. The problems of GaAs and other substrates are more severe.

The need for detailed understanding and control of impurities and defects will continue to grow in importance. Impurities and defects are both essential and detrimental to semiconductor devices, depending in part on their origin and control. Trace impurities, e.g., arsenic and boron, make silicon n-type or p-type, respectively. Electron conduction (n-type) and hole conduction (p-type) in adjacent regions form the basis for virtually all semiconductor devices. Unwanted trace impurities, e.g., sodium and other metallic ions, can ruin device stability or performance. Defects, often but not always associated with precipitation of some impurities, can cause catastrophic failure of circuits. On the other hand, in current silicon FET technology defects associated with precipitated oxygen impurities are useful for strengthening wafers and gettering other unwanted impurities from the crucial active regions of the processed silicon wafers.

Defects, especially those producing deep levels in the solid and the lattice relaxation around defects in the bulk, are understood in only a few cases. More must be learned about the chemistry and dynamics of defects and about the lattice distortions around defects at surfaces and interfaces.

Interfaces and Surfaces

In solid-state electronics each device is partitioned into two main regions: active regions in which energy is shifted from one frequency to another and contact regions that connect active regions to external terminals. Contacts invariably involve some type of interface between different materials. The active regions of the transistors used in an IC also involve interfaces: p-n junctions and Schottky barriers in bipolar and FET devices or the interfaces of the MOS structure itself.

As device sizes continue to shrink, it is evident that the surface-to-volume ratio of a single device is increasing at a rapid rate. Already in the case of modern VLSI transistor design, the transport in the device is dominated by the interface between the silicon and the SiO_2. A key question is the role of electrical contacts in submicrometer technology, not only from a fabrication and processing point of view but also from the standpoint of charge injection and extraction in these devices. Indeed, it is well accepted that the boundary conditions are of more than academic importance, even to the point of dominating the actual device behavior.

Schottky barriers appear as gate electrodes in metal-semiconductor FETs (MES-FETs) and as tunneling barriers in normal, so-called ohmic contacts. Yet after decades of study there is no first-principles theory of the Schottky barrier itself. While it is known that defects play a major role, there is no way to predict in advance what the barrier height of a new metal on a new semiconductor will be.

Silicon enjoys the advantage of having an excellent native oxide, perhaps the only such example among all possible semiconductors. Although nature was kind here, the details of how this oxide grows and why it is so good (in terms of the electrical properties of the interface) still elude a full understanding. Device progress has been remarkable without a proper theory of oxide growth and of interfacial electron states, but one can question whether this trend will continue. In other semiconductors, even less is known about the growth and interfacial properties of insulating layers.

These types of basic questions can be successfully addressed only through broadly based semiconductor physics and materials-science research programs. For example, the recent achievement of an atomically perfect (epitaxial) silicon-silicide interface, made possible by clever use of the newly developed MBE techniques, may well be a key step in solving the Schottky barrier problem.

Transport

There are several approaches to treating transport in semiconductor devices. The earliest treatments dealt primarily with the simple concepts of mobility and diffusion, although later approaches introduced a local electric field dependence in both of these quantities. Whereas such approaches will continue to work in silicon devices for several more generations of scaling, their inadequacy to treat higher-mobility materials such as GaAs is already known. The reason lies in the response time of the carriers to changes in the electric field. When these changes occur in a time comparable with typical relaxation times, local variables, such as mobility and diffusion, cannot be introduced. The situation will become even more complicated in future devices, which may incorporate heterojunctions and/or the so-called ballistic transport, where the conduction channel lengths are short enough that electrons can completely traverse it with only a few or no collisions.

Other short-channel effects include the spilling of charge carriers out of the channel into the adjacent insulating layer. Stability of the insulator-semiconductor interface and charge trapping in the insulator are related topics for silicon FETs and will be critical for related GaAs devices as well. The opposite extreme of long, narrow channels forms the basis of studies on reduced dimensionality and electron localization phenomena, currently of interest to research physicists.

We already know that modern devices exhibit quantum-mechanical effects. For instance, the electrons in the inversion layer of a silicon MOS-FET are quantized in the direction normal to the interface. These carriers then constitute a two-dimensional electron gas. The details of the scattering mechanisms and of the transport itself are more complicated in such a case. However, little has been done to incorporate quantum transport equations into device modeling. The case of transport normal to a heterojunction may be a better example. Modern technology can produce an interface between the two materials that is only one or two atomic layers in extent. Transport through such an interface has been couched in the normal, bulk semiconductor terms; little has been done to consider wave-function discontinuities, spatial variation of the scattering processes, and/or locally varying effective masses.

PACKAGING

High performance in computer systems was once determined by the capabilities of the individual chips, with the packaging together of these chips being a mundane consideration, required only to provide chip interconnections, input and output, power, and cooling. This packaging was technologically nonagressive and straightforward. Today chip circuit densities and speeds have advanced so far that the package performance has become one of the dominant performance limiters. Indeed, in the arena of very-high-performance supercomputers the package performance is becoming crucial to further advances. The intrinsic speed advantage of GaAs over silicon circuits greatly diminishes when systems of a large assembly of chips are considered. Some current estimates are that room temperature GaAs systems may have a performance advantage of only a factor of 2 (still significant) over silicon technology, in large part because of package considerations.

One example of a modern package component is a recently developed system referred to as a thermal cooled module, which provides mounting positions, interconnections, and cooling for more than 100 high-speed bipolar logic and memory chips. The module consists of 33 printed-circuit ceramic sheets laminated together with as many as 36,000 metal-filled holes to provide connections between the layers. The module provides transmission line connections between the chips, interconnecting 25,000 logic circuits and 65,000 memory bits. The use of this module reduced signal-delay time between chips by a factor of 4 compared with the previous approach.

A consequence of evolution in the importance and complexity of packaging is the need for increased understanding at the microscopic level of the interfaces between package structures. As a result, new opportunities are developing in microelectronics for interactions among physicists, materials scientists, and technologists in areas outside traditional semiconductor physics. Indeed, the emergence of packaging as a technical concern is focusing renewed attention on a variety of scientific problems—problems such as thermal transport across interfaces, adhesion, and production and processing of novel polymers and ceramic materials. Solutions of these problems can result both in the building of the fundamental knowledge base of condensed-matter physics and in the continued evolution of microelectronics.

As an example we can look at adhesion, a basic interface problem fundamental to the cohesiveness and permanence of packaging systems. The phenomena that cause adhesion at polymer/metal interfaces are only beginning to be understood in a fundamental way. Much more work must still be done to provide an understanding of the chemical and physical nature of these interfaces and the changes that can be made in order to enhance adhesion and prevent its deterioration under long-term thermal and environmental stresses. In a multilayer thin-film package that connects VLSI chips, thermal cycling creates mechanical stresses that act to delaminate the structure. In order to ensure integrity of the package it becomes crucial to know what the basic adhesive mechanisms between metals and polymers are and to propose new surface modifications that can hold together the thin layers in packages.

The problem of chip placement and wiring algorithms is another example that illustrates both how solutions can come unexpectedly from basic research and how essential it is to maintain strong interactions between science and technology. This case is a new application by solid-state physicists of statistical-mechanics techniques originally developed for magnetic-spin glasses and related problems in disordered solids. The researchers used the concept of simulated annealing, a mathematical construct, to provide procedures and criteria for the placement of circuit elements and integrated circuit chips within digital computers. By efficient computational techniques that simulate the slow freezing of a solid, they were able to reduce wiring congestion and thereby improve the layout of existing circuits. The new methodology also allowed previously intractable designs to be wired. The success of this physics model of circuit design gives one optimism that other physical models might be found for even more complicated automated design problems of VLSI.

MAGNETIC INFORMATION TECHNOLOGY— STORING THE BITS

An area critical to the computer revolution that has developed almost unnoticed when compared to microelectronics is magnetic

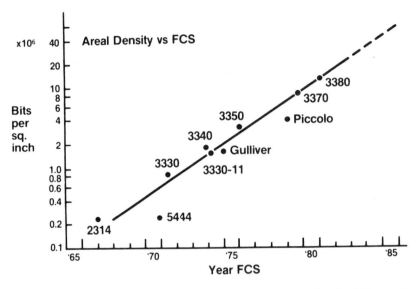

FIGURE 8.5   A plot of bit density versus first customer shipping date for rigid magnetic disk files.

information technology (MINT). MINT includes magnetic tape, floppy disks, and hard disks used to store information for computers. It also includes new technologies, such as magneto-optical recording and magnetic bubble technology, which may offer dramatic improvement in information storage density and reliability. The trend of increasing bit density for rigid magnetic disks is presented in Figure 8.5.

The rapid technological advances in this area have been critical to the computer revolution. Indeed, the dollar volume of magnetic storage sold for computers exceeds that of semiconductor memory devices.

MINT is based on physical principles researched decades ago. However, today MINT is in a state of rapid technological change and growth. This growth continues as computer users envisage applications that need even denser, cheaper, and faster magnetic storage devices. This demand has stimulated the invention of new materials and devices and the application of previously unapplied magnetic phenomena. It has also forced us to undertake demanding new research problems, such as finding ways to fly a recording head on an air cushion a distance above a disk surface comparable with the mean free path of molecules in air. Physicists, through their research and interaction with other scientists and technologists, are and will be instrumental in providing the knowledge that will make innovations possible.

Magnetic Recording

Magnetic recording is used for storage of both digital and analog information. Three forms of recording media are in common use today: tape, floppy disks, and rigid disks. Most commonly, to record and read information from these media, systems use recording heads that consist of a C-shaped soft magnetic material with a few turns of wire. For the highest possible recording density, the head must be brought as close as possible to the medium. With both tape and floppy disks, the head makes physical contact with the medium. With rigid disks, which spin at high speeds (the head-to-medium relative velocity is typically a few hundred miles per hour), the head is designed to float on an air bearing a few millionths of an inch above the rigid disk surface.

In magnetic recording, key performance parameters are storage density (the number of bits of information stored per unit area in the medium), access time, the data read rate, and reliability. The cost of a stored bit of information has been steadily decreasing almost inversely with storage density. Although during the past 20 years recording density has increased 1000 times, we are still orders of magnitude away from fundamental recording limits. Today a rigid disk of the most modern design stores 10,000 bits per inch along a recording track and has about 1000 tracks per inch, a recording density of 10 million bits per square inch. Recording densities as high as 200 times greater have been demonstrated in the laboratory. Even higher recording densities are theoretically possible.

Most recording media today consist of fine magnetic particles in a gluelike binder on the surface of a flexible (tape and floppy disk) or a rigid (rigid disk) substrate. The interaction of these particles in the coating and how the interactions affect the magnetic properties of the media are long-standing fundamental problems. Research is needed to provide the fundamental understanding of the relationship between the composition and microstructure of the particles and their magnetic properties.

In addition, there is a need for more research on continuous thin-film media made by vacuum deposition or electroplating techniques. Thin-film media have been used to demonstrate recording densities of 100,000 to 200,000 bits per inch; however, particulate media still dominate the marketplace. One of the problems is that the transition between oppositely recorded regions in a thin-film medium is not well understood or described. Irreproducibility in the recorded transition causes noise in the recording.

There is today a major unresolved controversy about whether

longitudinal recording (recording media magnetized in the plane of the medium) or vertical recording (recording media magnetized perpendicular to the plane of the medium) offers the highest density. Present understanding of the fundamentals of the recording process is insufficient to resolve even this basic question.

Further advances in magnetic recording are also fundamentally dependent on improvements in the magnetic head. Today there is great interest in recording heads fabricated with vacuum-deposited thin films rather than with bulk ferrite materials. These thin-film heads offer lower inductance (and therefore higher data rates) and higher potential bit densities than ferrite heads. The recent successful introduction of thin-film heads in rigid disk drives is a direct result of physicists' and engineers' working together to understand and control magnetic phenomena in NiFe films.

Now there is great interest in the potential application of amorphous alloys of magnetic materials in these thin-film heads. Amorphous magnetic alloys were invented in the 1960s and, in the thin-film form, were the subject of extensive research by physicists and materials scientists in the 1970s. Now the potential of these materials for application in magnetic heads is being actively investigated, but much remains to be done in understanding their properties before optimum materials for recording are achieved. One fundamental goal in such research is to develop head materials with larger saturation magnetization so that higher-coercivity media can be used.

Magnetic Bubble Technology

Magnetic bubble technology makes use of the same fabrication techniques as semiconductor ICs. It involves no moving parts, offers nonvolatile storage of information, has considerably higher storage density than semiconductor memories, and has proved to be extremely reliable in harsh environments. Current bubble devices in volume production offer storage capabilities of 1 Mbit/chip, but 4-Mbit/chip devices have been announced. Future higher-density devices will use new technology now being developed, such as ion-implanted contiguous-disk devices, and may offer on-chip current-access logic. The development of this technology requires research by physicists on the effects of ion implantation on garnet materials and on magnetic phenomena in these devices.

A new device technology, which has its origins in bubble technology but actually uses the structure in the domain walls of bubble domains

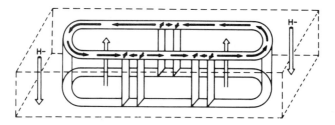

FIGURE 8.6   As described by H. Hayashi [*Hikkei Comput.* 97-102 (May 30, 1983)], Nippon Electric Company's Bloch line memory is based on the magnetic structure shown here. The magnetization in the wall of a stripe domain in bubble material can rotate in either of two senses from the direction of the field inside the stripe to the direction of the field outside the stripe. If a twist is put into the wall, the sense of rotation is reversed in a small portion of the wall, and a pair of Bloch lines forms—one on either side of the twist. The existence or absence of the Bloch line pair can be interpreted as a bit of information. Theoretically, this would permit information to be stored approximately 100 times more densely than in conventional bubble memories.

for information storage, has recently been promoted as offering storage densities beyond 10^9 bits/cm^2.

These new devices are called vertical Bloch line memories and are based on research in physics carried out in the United States a number of years ago—research that was terminated because no practical applications of the devices could be envisaged for the near future. A Japanese professor invented the device, and Japanese industry has been aggressively pursuing the technology. A description of the magnetic structure of a Bloch line memory is shown in Figure 8.6. If the United States is to have a position in this new technology, considerable work by physicists and engineers is required in the near future.

Magnetic-Optic Recording

Optical storage of information has dramatically entered the consumer marketplace in the form of video disks and compact audio disks. The first optical data recording products were introduced only recently. This technology provides storage of information at densities exceeding 10^8 bits per square inch, more than 10 times the density of present-generation rigid disks.

Although this technology offers high information storage density, it does not at present compete directly with magnetic recording because it is not erasable or rewritable. Computer systems today are designed around storage technologies that may be directly overwritten with new data.

Magneto-optic recording technology makes possible the high density of optical recording with the erasable and rewritable characteristics of conventional magnetic recording. In magneto-optic recording, the beam of light from a diode laser is focused onto the surface of a temperature-sensitive magnetic thin film. The local heating that is due to the energy from the laser beam causes a reduction in the magnetic field required to reverse the magnetization in the film, and, as a result, the magnetization in the heated region assumes the direction of the applied field.

To read the recorded information, the same diode laser is used, but with lower power output so that heating is insignificant. The plane-polarized light, on reflection from the film surface, suffers a rotation of the direction of polarization that is dependent on the direction of magnetization in the film. Thus, by using an analyzer and a photodiode, the magnetization direction in the disk, which is representative of binary data, may be read.

Amorphous thin-film materials are the present media for this technology and, being relatively new, are not well understood. Physicists are needed to help to develop a solid understanding of magneto-optic recording materials so that a high magneto-optical signal-to-noise ratio, good stability against changes in temperature, and a resistance to aging effects are achieved.

INTO THE FUTURE

If progress in microelectronics and magnetic information technology were to cease today, the effect of this technology on society would still continue to grow for decades to come. Circuit designers, computer scientists, the electronics industry, and society as a whole have just begun to realize the potential of what can be accomplished with today's technology. But progress in microelectronics is not going to cease today; this technology will continue to evolve and grow unabated for many years to come. We are far from reaching the foreseeable limits of what scientists and engineers can accomplish in this arena.

The next decade will see continued increases in VLSI circuit complexity, size, and density; decreases in circuit and system function cost; increases in device and system performance; and advances in packaging and development of the various materials, device, and circuit technologies. Microwave devices will continue to improve; monolithic integrated microwave circuits will be realized, with major impact on communications and other electronics applications. Magnetic information technology will continue to evolve, with storage becoming faster, cheaper, and denser. Today's supercomputer will become tomorrow's personal/laboratory computer. The effects of all this, when coupled with those of the emerging optical communications technology, will be remarkable.

As in the recent past, this continued technological evolution will draw broadly from the results of physics, applied physics, materials science, and electrical engineering research. Much of the focus will be on problems in well-defined areas such as those reviewed in this survey. But just as in the past, new technical developments, new solutions, and new advances will also come in unexpected ways from unexpected sources. A healthy, growing technology requires the continuation and expansion of fundamental research into all areas of knowledge, even those that may apparently affect this technology only remotely.

At the same time, the interaction between physics research and microelectronics technology will continue to be reciprocal. Physics will continue to benefit from the development of new materials growth, materials processing, and microfabrication techniques. From a more global perspective, the dependence of modern physics research on the utilization of high-performance electronics, microprocessors, and supercomputers will continue to grow even more pronounced.

What Next?

The evolution of existing microelectronics science and technology is relatively easy to predict over the short term. The performance of electronic devices and circuits will continue to evolve until it begins to approach the fundamental limits set by such barriers as the speed of light and the uncertainty principle. (As we have noted, some superconducting sensor devices are already essentially at their quantum limits.) The cost and capability of VLSI circuits will continue to improve.

But beyond these foreseeable evolutionary advances, important and profound as they will be, what are the prospects for future breakthroughs as revolutionary as the transistor? Of course, the invention of the transistor was almost unique in this century, comparable with the invention of fuel-powered machines of the previous century. But we note that in either case the real significance of epoch-making inventions is not that the inventions replace or do previously done functions better; it is in the new and unforeseen functions that they perform after people begin to understand and develop them. It is the low-cost, high-function IC of first ten, then a hundred, now thousands, and soon millions of transistors on one fingernail-sized chip of silicon that makes the transistor so important. While we can search for better ways to do or extend functions now done with microelectronics, we can surely expect that the real breakthroughs will bring now unknown new functions for electronic devices. The search for extended functions is likely to include three-dimensional, stacked circuits (present ICs are essentially planar with side-by-side transistors) for more function in less space, opto-electronic devices mixed with microelectronics on the same chip for alternative input and output, and self-organizing circuitry to overcome the ever-increasing complexity of logical and physical design. It is not hard to imagine that success in any of these related goals will provide more than just better versions of what we can extrapolate from today's microelectronics.

9

Applications of Physics to Optical Information Technologies

INTRODUCTION

Over the past decade the photon has begun to rival the electron in its impact on society and daily life. Ultimately the influence of physics advances in a variety of optical fields may surpass even that of the electronics technology that grew from the invention of the transistor more than 35 years ago. This chapter will describe briefly some of the major advances in physics research that have already found their way into optical technology and will illustrate the connections between fundamental research and applications that influence both quality of life and our national economy. In doing this we make some arbitrary limitations on the subject matter discussed. Optical physics embraces many applications. Some of them are already commercially important, but we have chosen not to dwell on them here. These include the commercialization of holography in supermarkets and elsewhere for pricing and inventory control; the widespread use of lasers for monitoring, control, and processing in construction and manufacture as well as in sophisticated automation systems; lasers in surgery; sophisticated techniques for photography from satellites and high-altitude aircraft; and the important science and industry of photocopying and image reproduction.

We focus on *optical information technologies* because of their potential societal impact and because they now stand at a particularly attractive stage for illustrating the process of innovation.

161

Over the coming decades optical technologies in synergy with electronics promise to alter substantially major aspects of life in modern society. Low-cost transmission of massive information flows on optical fiber together with ever cheaper and more efficient information processing will help to bring about entirely new options for interactions within our society. Networks of powerful, inexpensive, user-friendly computers (whose insides will combine electronic wizardry with blinding optical speeds) promise to put each citizen in full, real-time, switched, color video contact with any other citizen, expert, data base, business, or library.

Options available to the average person for education and entertainment, for communicating instead of commuting, and for teleconferencing in place of traveling will multiply exponentially as a result of optical information technologies. Monitoring and metering functions in the home as well as the factory will be done remotely, with greater precision, and at lower cost than is possible today. The fabrication and flow of goods throughout society will also be greatly enhanced.

The symbiosis between optical and electronic technologies will continue to expand. The resulting network will increase the power and productivity of every element in that society. Each person's expertise will be multiplied by unencumbered access to all the knowledge and insight of every other expert. Because future remote encounters will be easy and intimate, they will become qualitatively more acceptable, more personal, making travel and physical presence truly optional for the first time in human history.

We divide the remainder of this chapter into three sections. The first, on optical communications technology, elucidates the role of physics in optical communications systems (sources, detectors, the light-guide fiber, and overall systems configurations). The next section treats optical information processing, ranging from today's video disk to the recently discovered optical phenomena that offer the promise for *all-optical* processing systems of greatly increased speed, flexibility, and information density. The final section, on the photonic future, discusses other active research areas likely to lead to further significant advances in optical technology. For example, the relentless drive of physics toward nature's fundamental limits has recently led to the generation of optical pulses lasting only a few quadrillionths of a second. The prospect for integrated optics in the 1990s analogous to integrated electronics of the 1960s and 1970s appears secure as a result of recent advances in materials and optical physics. Exploitation of the switching speeds offered by optical phenomena may lead to completely new strategies for high-speed computing.

This chapter thus illustrates the continuum joining fundamental research, technological application, and economic and societal impact embodied in the science and technology of optical physics.

OPTICAL COMMUNICATIONS TECHNOLOGY

Human beings receive most of their information about the universe from photons. Beginning with the information directly perceived visually, our perceptions embrace as well detailed fundamental information about atoms, molecules, the structure and dynamics of matter, and the cosmos itself revealed by photonic spectroscopy. Only in the past half-decade have practical optical communications systems added an important new dimension to humanity's optical information channels. The quarter-million miles of optical fiber deployed in commercial optical communications systems last year represent only the leading edge of an explosion in worldwide optical transmission capabilities. Over the next few years systems with incredibly greater information-carrying capacity and span distances will become commonplace. These will help to make accessible to citizens throughout the world forms and amounts of information that will transform their ways of life. It is instructive to understand the role of physics in driving and sustaining this revolution.

Optical Communications System Components

Optical communication certainly dates from at least the first prehistoric signal fires used to communicate between hilltops. Even then the essential elements of an optical communications system were evident: a source of an optical signal, an appropriate medium through which that signal can propagate, and an appropriate scheme for detecting the signal and extracting the information therefrom. Over the centuries, the distances and information bit rates embodied in primitive signal-fire technology proved inferior to other, primarily electrical, transmission schemes. As society has grown more complex the need for information transmission capacity has expanded relentlessly, as shown in Figure 9.1. Indeed, over the past century the demand has increased exponentially, doubling every 5 years. While it has long been realized that the higher frequencies in the infrared or optical regimes offered enormous increases in potential bandwidth available for information transmission, until quite recently neither the means nor the need for such dramatically increased capacity existed.

The development of optical communications has been driven largely

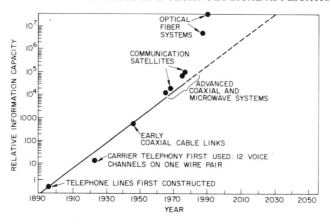

FIGURE 9.1   Historic growth in telecommunication capacity.

by the electronics revolution (from the transistor through integrated circuits to the proliferation of smaller and faster computers) through the prodigious growth in demand for information transmission capabilities that it has generated. Indeed, without that demand the crucial physical elements of today's optical communications systems may have remained mere laboratory curiosities.

These elements all trace their origins to physics: the invention of the laser; advances in semiconductor physics permitting fast and sensitive optical detectors; the fundamental understanding of propagation effects and guided waves; and insights into the optical, mechanical, and chemical properties of glasses and amorphous solids.

Early lasers based on ruby rods or gas plasmas were far too expensive, unreliable, and bulky for deployment in a practical communications system. Several years of research in optical and solid-state physics were required before the first semiconductor laser was demonstrated in the middle 1960s, and several more before long-lived, reliable, continuous room-temperature operation was achieved in 1973. Crucial to these developments were the understanding of optical properties of semiconductors, the role of energy-band structures, the nature and behavior of dopants, the sources of carrier lifetimes and optical thresholds—all vigorous physics research activities in the 1960s. Today's communication lasers are salt-grain sized, ultrareliable, rugged, and highly controllable. A typical semiconductor laser is diagrammed in Figure 9.2. These lasers represent the culmination of diverse research programs from decades past.

Semiconductor physics research has been essential to the evolution of optical device technology. For example, the electronic energy-band structure of a semiconductor determines the optical device uses to which it might be put. Electrons may be excited into the higher-energy conduction band by the application of voltage across a *p-n* junction, leaving behind holes (positively charged carriers) in the valence band. These electron-hole pairs subsequently recombine with the associated emission of photons. By proper design of current and light-confinement regions within the semiconductor material, under even relatively modest total currents the electron-hole recombinations will cause stimulated photon emission and, thereby, laser action. Lower current densities cause the spontaneous emission operative in light-emitting diodes (LEDs). The emitted photon energy (or light wavelength) is directly determined by the energy-band structure, which in turn depends on the semiconductors' crystal structure and chemical composition. The same phenomena, run in reverse sequence, are utilized in photodetection, so that the materials and designs of today's optical sources and detectors share considerable commonality. Concepts,

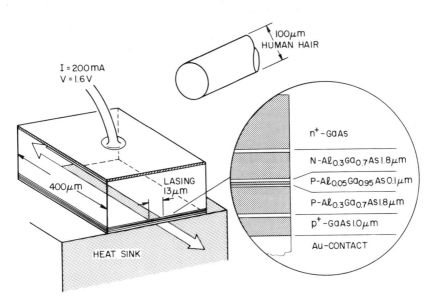

FIGURE 9.2   Schematic of compound semiconductor laser driven by electrical current flow through the *p-n* junction. Output wavelength of 0.82 μm is determined by the chemical composition of the active P-Al$_{0.05}$Ga$_{0.95}$As layer.

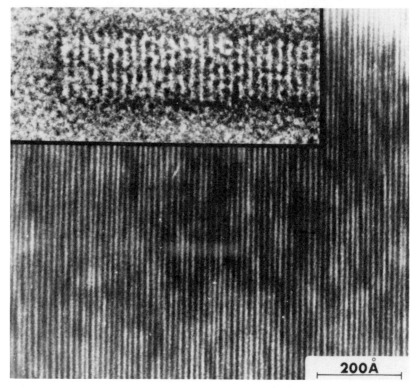

FIGURE 9.3   Electron microscopy has been used to reveal the regular composition variations produced by MBE. Growth direction is from left to right. Dark bands are GaAs, and light bands are GaAlAs. The typical layer thickness is only about 7 Å, or four atomic planes. Insert shows an electron micrograph of a virus molecule at the same magnification.

parameters, and viewpoints of concern only to the research physicist two decades ago have today become the tools of the optical systems engineer.

Continuing research in physics will soon free engineers from the restrictions to pure bulk semiconductors; from it has come a new field, which we might call band-gap engineering. For example, by a technique called molecular-beam epitaxy (MBE) precisely controlled streams of selected atoms are played across the surface of a perfect semiconductor crystal. Control is so good that new materials (of structures, compositions, and properties never envisioned in nature) may be built up atomic layer by atomic layer, as illustrated in Figure 9.3. The resulting energy-band structure can thus be made to precise specifications. This precise controllability of the operating wavelength in a communications system is essential because the transmission medium (glass light guide) itself exhibits strong wavelength dependence of both its transparency and its optical bandwidth. MBE materials have so far been fabricated primarily into exotic, experimental optical sources and detectors in the research laboratories, but they will surely reach widespread technological application fairly soon.

Physics research has also played a fundamental role in the development of the glass-fiber light-guide transmission medium. Serious consideration of light-wave transmission on optical fiber for communications systems dates back to the mid-1960s. Conceptually it built on a substantial experience in guided-wave transmission physics of microwave/coaxial systems. Waveguide phenomena, quantified through solutions to Maxwell's equations under the appropriate boundary conditions, are common features of optical light guides and their much longer wavelength microwave counterparts. The phenomenon of light guiding itself is based on an even older concept in physics—that of total internal reflection, discovered in the mid-nineteenth century. When light rays strike an interface between two media with different refractive indices, they may be reflected without loss, provided that certain simple relationships among the refractive indices and the ray's angle are satisfied. By 1970 applications of electromagnetic propagation theory had quantified the optical and geometrical properties necessary for high-efficiency, high-bandwidth optical-fiber light guides. The modal properties of light guides were elucidated, and both single-mode step index (shown in Figure 9.4) and multimode graded-index light-guide designs were identified as promising routes to very high (several hundred megahertz) bandwidth transmission.

Also quantified were the harmful effects of even the most minute variations in light-guide diameter. In practical terms these translate into requiring that over distances of many kilometers the light-guide diameter not vary by more than one one-hundredth the diameter of a human hair. Early systems made extensive use of graded-index multimode fiber, particularly for short-distance applications. As we look ahead, emerging applications will make increasing use of single-mode-fiber technology. Physics has continued to play a key role in designing ever more sophisticated single-mode light guides capable of transmitting at multigigabit rates over distances exceeding a hundred miles and operating simultaneously on several wavelengths. For example, one contemporary experimental light-guide design should permit operation over the entire wavelength band between 1.3 and 1.6 μm even at these ultrahigh bit rates.

Before such light-guide design considerations could become of practical concern, the major longstanding impediment to long-distance

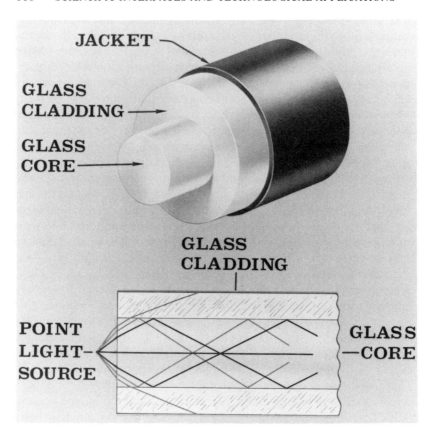

FIGURE 9.4   To produce reliable internal reflections, as shown in the lower diagram, high standards of optical and mechanical perfection must be maintained in fiber production.

optical transmission—optical loss—had to be removed. Physicists who have studied the optical properties of condensed matter for several decades might well be surprised at the applications of their discoveries being made today. The basic phenomena of infrared absorption, ultraviolet absorption, and light scattering from molecular vibrations and density and concentration fluctuations determine the intrinsic transparency of a pure material and serve to define the wavelength windows for transmission. When all impurities are absent, a material's optical transmission properties are dominated by Rayleigh scattering (which decreases at longer wavelengths) and by fundamental infrared absorption of the material's molecules (which increases as the wavelength increases). In fused silica (the basic component of today's finest fibers) the combination of these two effects provides a loss minimum of approximately 0.15 dB/km at wavelengths near 1.5 μm. Changing the wavelength away from this value increases the optical loss.

In addition, the wealth of spectroscopic information and impurity-specific data amassed by physicists interested in such phenomena as defects and color centers proved invaluable in identifying and quantifying loss-producing impurities, as seen in Figure 9.5. The interplay among physics, materials science, and chemistry in the area of optical-fiber light guides is a powerful illustration of the societal value of interdisciplinary applied research.

From Figure 9.6, it can be seen that as recently as the late 1960s the lowest loss attainable in a research-grade optical fiber was several hundred decibels per kilometer. Today, in mass production, losses of 0.25 dB/km are routinely achieved—an astounding improvement of more than a thousandfold. This dramatic reduction in optical loss has brought optical communications from the status of an experimental curiosity to a multibillion-dollar set of businesses in just over a decade. Although physics provided much of the basic understanding of the necessary optics and materials properties, much of the success in translating this understanding into a practical technology is due to chemists, materials scientists, and chemical engineers. To achieve the

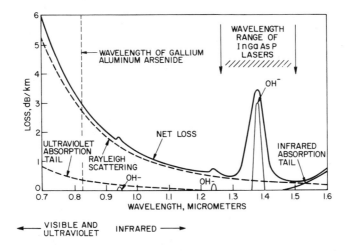

FIGURE 9.5   The ultimate transparency of glass fibers depends on the wavelength of several fundamental mechanisms. Except for the impurity absorptions (labeled OH⁻) that are due to small amounts (less than 30 parts per billion) of fragmentary water, the losses shown above are intrinsic for silica-type glasses. Metallic impurities (such as Fe and Cu) have been reduced below the parts-per-billion level.

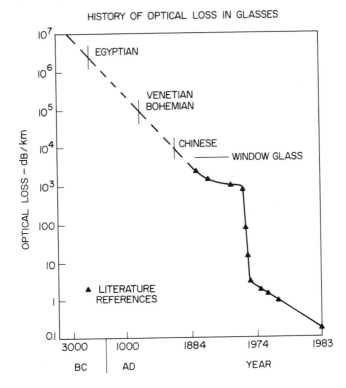

FIGURE 9.6   The concerted efforts of research physicists and chemists have brought about astounding reduction in optical losses. Today's best fibers are several thousand times more transparent than those of a relatively few years ago—and they now approach the fundamental limits allowed by the pure SiO_2 material itself. Such fibers can readily transmit fairly weak optical signals for more than a hundred miles without amplification.

parts-per-billion impurity levels necessary for low-loss light guides, vapor-phase chemical processing techniques—completely novel from the point of view of traditional glass fabrication—had to be developed. At least three of these are today enjoying significant commercial implementation. For the next generation, one of the most promising new methods uses a plasma fireball stably suspended within the flow of reacting chemicals inside a rotating glass tube to deposit microscopic glass particles with high efficiency. Thus, somewhat unexpectedly, even plasma physics may soon make essential contributions to economic large-scale manufacture of optical communications fiber.

Evolving Systems Configurations

As optical technology expands, its reliance on the discoveries of physics will continue to diversify. Early light-wave technology utilized multimode fibers and GaAlAs optical sources (lasers or LEDs) operating below 0.9-μm wavelengths. Bit rates were measured in tens of megabits per second, and distances in the few-kilometer range. On the foreseeable horizon, two distinct frontiers emerge—each with its specific challenges. On the one hand, light-wave systems for short-distance, relatively low-bit-rate applications such as computer interconnections, local area networks, and on-premises uses imply lower-cost equipment and the acceptability of relatively modest performance levels. Easily joined, low-cost, moderate-loss fibers and very-low-cost sources and detectors of only moderate speed and sensitivity will be required. On the other hand, systems for long-distance, high-bit-rate applications such as undersea and extended terrestrial networks place a premium on performance, driving us to maximize the distance between repeaters and the number of bits per second transmitted over a single fiber. The high-performance systems appear to offer greater opportunities for continued novelty in applying recent physics discoveries (sources, detectors, fibers, and electronics).

Technological Challenges

While systems economics help to determine future systems configurations, successful resolution of several technical challenges is equally important. These include increased source spectral purity and higher modulation bit rates, lower-loss fiber materials, novel broad-band light-guide designs, unusual modes of signal propagation, and more-sensitive signal detection.

The fabrication of precisely controlled single-frequency miniature semiconductor lasers that produce several billion pulses per second presents a formidable challenge. Physicists have recently understood more clearly the fundamental conflicts between the objectives of high-speed pulsing and spectral purity of laser output. The statistical physics of the transient photon amplification process as well as the influence of the changing electron density on the laser frequency may limit the pulse rates from semiconductor lasers to less than ten billion per second. Although this limitation may appear to be of only academic interest, it represents a bit rate several thousand times *less* than the shortest light pulses ever generated could in principle permit. Research on new laser designs may overcome these obstacles. The distributed-feedback laser represents an ultrastable single-frequency device, and a variety of external coupled-cavity laser designs (in which a control section can be appended on a microscopic scale to the active section of a miniature semiconductor laser) show great promise for controlled tunability and frequency keying. Using such lasers, recent systems experiments demonstrated unrepeated transmission over distances in excess of 100 miles at bit rates in excess of several billion per second. New bit-rate/distance records are being set nearly every day and are now recognized as the figure of merit for optical communications, in analogy to the transistor density for electronics.

Research on other novel characteristics of fibers may lead to practical exploitation of the phase coherence of laser light in communications systems. Today's systems operate in a digital, pulsed format wherein light pulses are rapidly turned on and off and the information encoded in the resulting sequence is directly detected downstream. Noise associated with the detection scheme ultimately limits the length of such a system. *Coherent detection*, in which the incoming optical signal is detected by beating with a local optical oscillator (as is now done at much lower frequencies in FM radio), should provide an easier path to ultimate sensitivity than does direct detection. To preserve the phase and the state of polarization over tens of kilometers requires still more research into light-guide design and fabrication methods, but polarization-preserving fibers have already been demonstrated.

Fiber-Optic Sensor Technology

An optical fiber, by virtue of its ability to concentrate light over long distances, permits unprecedented integration over those distances of the accumulated effects of the fiber environment on the state of light passing through it. Exquisitely sensitive devices have already been constructed based on the optical phenomenon of interference using the fiber medium to express changes in temperature, pressure, magnetic field, or rotation speed. Phase shifts of a small wavelength fraction over fiber distances of hundreds of feet can be easily detected, providing increases in measurement sensitivity that exceed 10 millionfold.

By exploiting the dependence on external parameters of the glass fiber's refractive index, absorption, or birefringence, the nucleus of a new technology has been formed. Refinements now include the use of magnetostrictive, electrostrictive, or thermostrictive fiber coatings and the imaginative fabrication and exploitation of cylindrically asymmetric, birefringent fibers.

The use of fibers in sensors, experimentally embodied today as fiber interferometry and fiber holography, will surely be of increasing technical, commercial, and military significance in the coming years.

Other modes of optical information transfer, such as fiber bundles for direct image transmission, are also under investigation. Image-transmitting fiber bundles with nearly 100,000 pixels have been fabricated. However, neither the transparency nor the resolution is yet sufficient to permit use of fiber bundles in more than a few specialized applications. Fibers may also be used for remote holography, opening the possibility of long-distance transmission of three-dimensional images.

OPTICAL INFORMATION PROCESSING

As was discussed in Chapter 8, attainable electronic processing power per unit area or per unit cost has continually increased, essentially doubling every year for the past two decades. Currently the most sophisticated production electronic device incorporates into a single semiconductor chip more than a million memory elements. In the laboratory that number has reached well beyond a million, and we are still quite some distance from the fundamental limits dictated by physics. Nevertheless, it is appropriate to ask whether or for which applications electronic information processing will continue to be the strategy of choice.

Research in optical and materials physics has raised the possibility of all-optical information generation, storage, switching, and computation. Surely for certain tasks the extreme compactness and serial processing architectures available and foreseeable from electronics technology will continue to dominate. But for others, for which speed and power dissipation are more important than size, or for which the format or the sheer amounts of information to be processed favor parallel architectures, *all-optical information processing* holds considerable promise. Although it is premature to predict the form or speed of this evolution, we are already in a position to point out several of the key elements. These include optical storage techniques; means for performing logic operations on optical signals both in the hybrid electro-optic, magneto-optic, and acousto-optic technologies as well as in integrated optics; and optical nonlinear effects on which optical computers might be based.

Optical Memory

The most familiar format for optical storage and retrieval of information is the video disk, which a sequence of bits is written onto a master rotating disk at densities exceeding several million per square

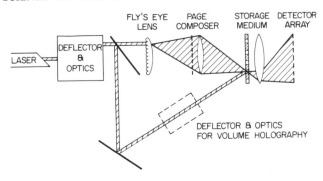

FIGURE 9.7   Holograms are exposed by the interference of two coherent light beams, one acting as a reference, the other carrying the object's information. A simple extension to three dimensions may be effected by stacking the storage-media plates.

centimeter. The disk's reflectivity or transparency is modified—typically by the material's thermal response to tightly focused pulses of light. As with audio disks, video disks are capable of being mass copied, but the master disks can be exposed only with rather sophisticated equipment. Memories with repeated read-write capabilities are much more difficult to engineer than are write-once, read-only memories. The trade-offs between permanent storage and erasability in optical disk technology have so far fallen in favor of the archival materials and techniques. Because the light can be focused to a 1-μm-sized spot, bit-storage densities of nearly one billion per square inch can be achieved. Already, cheap and reliable miniature semiconductor lasers are employed in commercial video disk players and are having an impact in the entertainment industry.

Current optical storage techniques are essentially two dimensional. With the restriction to two dimensions, holographic memories are capable of effective storage densities comparable only with those of video disk technology. However, holography offers the added ability to store information at different depths in volume formats so that bit densities as high as a trillion bits per cubic centimeter in stacked holographic arrays or volume holograms might be achieved. A read-write setup for a two-dimensional holographic memory is shown in Figure 9.7.

The physical mechanisms underlying various optical memory schemes have emerged quite directly from physics research on the optical properties of materials. These fall into several classes. One is the essentially thermal effects, such as ablation, the crystalline-to-amorphous transformation, and thin-film distortion or local melting.

Another is optically induced changes in the material's absorption, emission, or refractive index. In photochromic materials optical radiation changes the local optical absorption of the material either at the exposure wavelength or at some other (reading) wavelength. This has been demonstrated in alkali halides, glasses, and various organic materials using both linear and nonlinear optical effects. Practical read-write memories based on magneto-optical effects have also begun to appear.

A phenomenon that does not directly utilize change in absorption is the photorefractive effect, in which exposure causes a change in local refractive index of the medium. Direct photorefractive effects change the electronic state of a material by direct optical transition. Indirect photorefractive effects involve local heating and consequent configurational changes in the material. Again, these effects may be either linear or nonlinear, involving single- or multiple-photon exposure. Response and relaxation times as short as a few picoseconds and as long as many years have been observed. General trade-offs must be made among response time, storage capability, and read-write sensitivity offered by different materials. A particularly promising new direction is the photorefractive effect in doped semiconductors in which the decay time, sensitivity, and operating wavelength may be controlled over rather wide ranges by choice and concentration of dopants.

A second exciting variation on optical storage is based on a phenomenon called spectral hole burning. In certain materials optical absorption lines associated with impurity ions are inhomogeneously broadened by the random environment in which the absorbing ion finds itself. Strong illumination by narrow-band light falling within such broad absorption bands can bleach out the absorption over the same narrow frequency band. By bleaching different spatial configurations with several different narrow-band illuminations, storage of many holograms within the *same* spatial volume may be achieved. These may be separately read without interference, by light of the appropriate frequencies. Thus it may be possible to increase substantially the $10^{12}/cm^3$ bit density achieved in a volume hologram that is sensitive at only one wavelength. Several hundred trillion bits may be stored in a sugar-cube-sized material. This corresponds to several *million encyclopedia volumes*. Such memories place severe demands on the spectral purity and stability of laser frequencies, and the relaxation times and the durability of spectral-hole-burned optical memories will require more research and development, but the amounts of information that such memories may handle are truly prodigious. Storing the entire contents of the U.S. Library of Congress in a few cubic centimeters of material is conceivable. Furthermore, such memories can be reconfigured quickly by proper sequencing of writing and bleaching optical pulses and thus could form the basis for parallel-processing logic machines as well.

Integrated Optics and Optoelectronics

The integration of several optical and electronic functions within a single device is often called integrated optics, and its realization has been a dream of engineers for many years. The basic elements have been demonstrated in the laboratory, but practical implementation to date has embraced only a few simple hybrid devices such as electro-optic modulators and couplers. On electronic command, these switch optical information incident in one or more channels into one or more different emerging optical channels. Arrays of such switches can add, monitor, manipulate, and read information onto or from an optical beam. Such optoelectronic devices permit extremely high-speed modulation (gigahertz rates and beyond) and processing as well as broad-band switching of digital information in the optical format. Similar hybrid devices based on acousto-optic and magneto-optic interactions may find relatively near-term use in specialized applications. A typical example of an integrated optical circuit is given in Figure 9.8.

Today's complicated optical regenerators require separate electronic circuits for the control of both the optical detector and the optical source. Research into integrating these separate optical and electronic functions has made considerable progress in recent years. Amplifiers and optical detectors, lasers, couplers, modulators, and control units have been integrated into the same chip of a III-V semiconductor, for example. Arrays of lasers and/or detectors have also been fabricated within a single chip and offer considerable promise for lower-cost, higher-reliability optical processing devices in the relatively near future.

THE PHOTONIC FUTURE: TODAY'S RESEARCH FOR TOMORROW'S TECHNOLOGY

All-Optical Logic

Optical signals may in principle be manipulated and processed at much higher speeds than electronic signals simply because the mini-

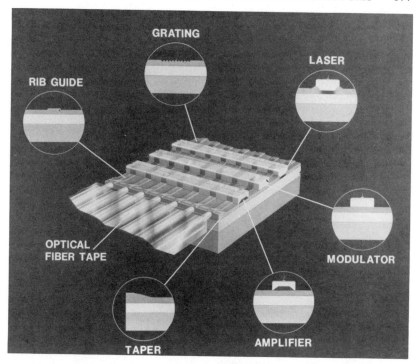

FIGURE 9.8 The increasingly complex integration of optical and electronic functions onto a single semiconductor chip promises to reduce the size and cost and to improve the reliability of high-speed optical information systems. Medium-scale integrated optical circuits like this one are already being studied in physics and engineering research laboratories.

mum duration of available optical pulses (10^{-14} s) is so much smaller than that of electronic pulses. However, the very weakness of the photon-photon interaction (nonlinear optical coefficients) that makes possible long-distance transmission of light through materials makes it difficult to modulate information in one optical channel directly by imposing an optical signal in another channel. Typical optical nonlinearities require interaction lengths of several millimeters for such modulation (compared with tenths of micrometers for electrons). Unless means can be found to enhance these nonlinear interactions greatly, purely optical logic devices will never be competitive with electronic devices in terms of size. Thus, even though nonlinear optical interactions have been shown capable of performing all the basic logic functions (AND, OR, NAND, NOR,. . .), an all-optical computer based on more-or-less straightforward analogy with its electronic predecessor would occupy considerable space.

There are two, not necessarily mutually exclusive, approaches to this challenge under current investigation. One is effectively to enhance the optical nonlinearities by several orders of magnitude, so as to bring today's millimeter optical length scales closer to the micrometer sizes characteristic of electronic logic gates. Another is to exploit the inherent parallelism of optical processes in totally new computer architectures. One means of enhancing optical nonlinearities is to make use of resonant optical phenomena and guided structures. Two-dimensional arrays of wavelength-sized metallic ellipsoids have been shown to enhance the local optical field at least a hundred times. This enhancement—discovered in the fundamental research on surface-enhanced Raman scattering—has already been proved useful in more efficient fluorescence and harmonic generation from thin layers, surfaces, and interfaces. The challenge to effect similar increases in three-dimensional or volume configurations remains a subject of active research.

A completely new phenomenon—optical bistability—discovered in 1976, has already been exploited to fashion miniature optical analogs of the electronic transistor. It is perhaps most clearly illustrated by considering a resonant optical structure such as a Fabry-Perot cavity in which a nonlinear optical material is placed, as presented in Figure 9.9(a). The resonance properties of the cavity are (by virtue of the nonlinear materials) strongly dependent on the frequency and intensity of the light incident upon the cavity. The cavity may be considered a periodic passband filter whose transmission at a given wavelength depends on the optical status of the material within the resonator. If this status is sensitively dependent on the intensity of the light present, the possibility of bistable operation exists. If the total intensity of the light in the resonator is due to two sources, a signal beam and a control beam, varying the intensity in the control beam can control the transmission or reflection of the signal beam quite accurately. The device may be simply regarded as an optically activated optical switch [Figure 9.9(b)]. Variations on this theme permit implementation of any basic logical operation.

The sensitivities and response times of optical bistable devices have improved rapidly. For example, working at wavelengths close to the band gap of a semiconductor such as GaAs greatly enhances the optical nonlinearities, so the optical power required for switching is correspondingly decreased. Advances in materials physics have permitted

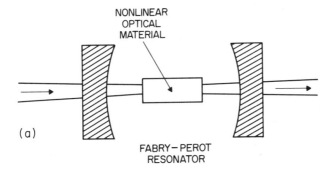

(a)

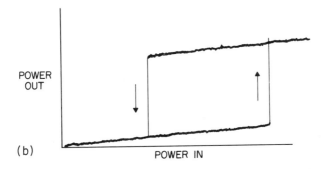

(b)

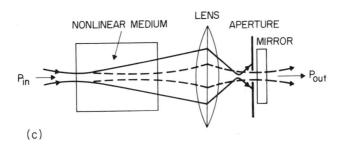

(c)

FIGURE 9.9 Optical bistable devices exhibit input-output characteristics of the same form as the more familiar electronic switching elements. However, they require no external electrical signal and may be switched on in trillionths of a second. For ultimate miniaturization and low switching powers, resonant structures appear most promising. However, the simpler, larger, and slower nonresonant versions may also prove useful.

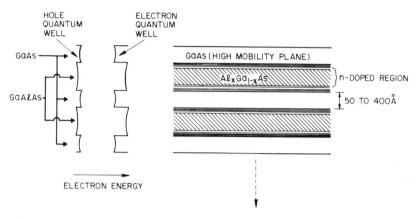

FIGURE 9.10   The composition control provided by MBE permits the formation of novel quantum states for the electrons in a semiconductor. The size and composition of such multiquantum wells can be used to sharpen and enhance the material's optical response as well, thereby improving performance of optical bistable devices.

the fabrication of multiquantum-well structures in the III-V semiconductor family, wherein the sharp and extremely strong exciton adsorption resonances in the vicinity of the semiconductor band gap may be precisely tuned and controlled. An example of a modulation-doped quantum-well structure is given in Figure 9.10. This represents a physical realization of electronic energy-band structures that until a few years ago appeared only in quantum-mechanics textbooks. The ability to control and fabricate essentially perfectly controlled heterogeneous crystals on an atom-by-atom basis lies behind this advance. Operating a bistable device based on a multiquantum-well exciton resonance has further reduced power requirements for optical switching.

Optical bistable devices can of course function as an active memory as well as logic elements. They also permit mixing and modulation of information on optical beams of different wavelengths so that their place in wavelength-multiplexed optical transmission systems may emerge.

The second major research direction in optical logic lies in parallel processing. So far we have discussed optical processing of information in strict analogy with electronic processors. That is, the architecture that was implicitly assumed has been based on serial processing of information. Under this restriction the speed advantage offered by optical techniques will be largely offset by their size disadvantages. To take full advantage of the optical processor one must consider parallel-processing architectures. It is quite possible to envisage several million bistable elements per square centimeter, defined and reconfigured by the array of optical signals incident, for example, upon a III-V semiconductor slab acting as a Fabry-Perot resonator. Depending on the dopants used and on the operating wavelengths, such an array could respond or relax on time scales ranging from trillionths to thousandths of a second. At a characteristic time of one billionth of a second it could execute nearly a *quadrillion logical operations per second* (orders of magnitude beyond the capabilities of today's best electronic computers). Of course the control and flow of the information in so many parallel streams is a formidable challenge indeed and is a long way from realization. Mundane considerations such as heat dissipation could substantially reduce the practically achievable bit density and processing rate. Nevertheless, the parallel format is ideally suited to the processing of images, a format that is inefficiently handled by sequential processors.

The rate at which advances in materials and optical physics are being married to engineering and computer design considerations identifies optical information processing as one of the most exciting technologies on the horizon today. A schematic of an optical array processor is shown in Figure 9.11.

Slicing the Second: Fundamental Limits to High Speed

Optical physicists have within the past 2 years greatly extended the world's record for the fastest laser pulses—down to a duration of 90 quadrillionths of a second—by using a novel dye-laser scheme called colliding-pulse mode locking. These pulses have been even further compressed (to about 10 quadrillionths of a second) by running them through an optical-fiber/grating combination. These pulses are so short that their light (which travels at 186,000 miles per second) has time only to span about one twentieth the diameter of a human hair. There is new physics even in the pulse-compression scheme. It may be the first practical use of self-phase modulation—an exotic nonlinear optical effect present in all materials. Colliding-pulse mode-locked (CPM) lasers today are large, expensive, and temperamental. It is unlikely that they will ever be directly miniaturized to the degree of today's relatively simple semiconductor lasers, but their increasing availability in research laboratories has opened to study entire new classes of

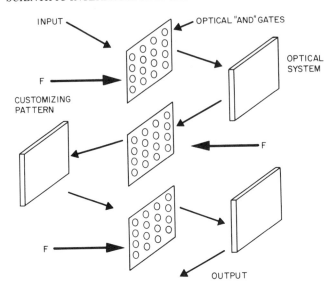

FIGURE 9.11   Illumination of entire arrays of optical logic gates simultaneously with fixed or transient customizing patterns permits not only the parallel handling of many information channels but optical reprogramming at rapid rates as well.

ultrafast phenomena in atoms, molecules, chemical processes, and electronic materials. From the discoveries that CPM lasers will make possible, other yet unanticipated ultrafast light sources are sure to emerge.

Exotic Propagation Modes and Media

The nonlinear optical properties of materials are being combined with sophisticated wave-propagation theories to open other prospects for high-speed information handling. A new kind of light pulse—the soliton—was propagated in optical fibers for the first time in 1981. First observed in water channels in the nineteenth century by a British scientist, such stable, finite-amplitude solitary waves remained a curiosity for many decades. During the 1960s plasma physicists refined the soliton concepts, and over the intervening years a wide variety of strongly nonlinear, dispersive physical systems have been shown to sustain solitons. However, optical fibers have provided the opportunity not only to study the formation, evolution, and propagation of optical solitons over long distances but also to amplify and regenerate them in novel ways. During the past year, the first soliton laser was demonstrated. It provides a new kind of uniquely controllable light source. Soliton formation can easily shorten the light-pulse output of a conventional solid-state laser to the subpicosecond (less than a trillionth of a second) regime.

Soliton pulses are self-shaping in that the total pulse energy determines the pulse duration. This unique property may someday be exploited in optical transmission systems wherein pulse reamplification and reshaping would be done entirely without the need for electronics. Solitons may be controllably reamplified merely by the injection of another light signal into the fiber, for instance, through such nonlinear optical processes as the stimulated Raman effect or four-wave mixing.

Research on new fiber materials is growing rapidly throughout the world. In fused silica (today's fiber material) the total intrinsic loss resulting from scattering and molecular infrared absorptions reaches a minimum value of 0.15 dB/km at a wavelength of 1.55 μm. New materials of heavier, less polarizable, and/or less tightly bound atoms could have intrinsic losses well below 0.15 dB/km—with obvious consequences for long-distance commercial or military communications. Such materials might also transmit at wavelengths substantially beyond 2 μm (where silica is essentially opaque), offering prospects for remote thermal sensing, surgery, machining, or perhaps even infrared power transmission. All these goals require materials of substantially higher purity than has been achieved thus far. Concerted research in the physics and chemistry of optical materials and processes must be continued if these potentials are to be realized.

Synergy Between Optical Science and Fundamental Physics

Finally, many of the same optically related research programs that may lead to future applications are also discovering phenomena of fundamental importance to other branches of physics research. To the list of solitons and ultrashort laser pulses should be added transient coherent phenomena, such as photon echoes recently observed in specially doped glasses at low temperatures and the increasingly complex sequences of bifurcations in optical bistable systems. The former offers a probe of unprecedented sensitivity for determining the energetics of individual atoms or molecules within a random environment. It will help us to understand the nature of spatial disorder and the localization of energy. The latter provides new insights into the evolution of turbulence—a mysterious state in which chaos in both space and time exists. Traditionally viewed from the perspective of fluid flows, turbulence has now become widely regarded as a general phenomenon whose development may obey quite similar but previously unappreciated "complexifying" sequences of transitions embodied in vastly different kinds of physical systems. Optical experiments on multistability bring unique precision and parameter variety to this important class of problems.

The research areas discussed in this final section vary greatly in their predictable prospects for application and for societal or economic impact. And, of course, they do not begin to exhaust the wide range of physics research from which future advances in optical technology may emerge. Just as today's soliton light pulses trace their origins to water flows first observed a century ago, the genesis of some important concepts for future optical technology may reside in quite different research activities from those we have discussed here. But if history and experience have any indicative validity, we may be confident that the clearly demonstrable threads connecting fundamental research, application, and societal impact that have characterized optical technology in the immediate past will continue to do so in the future.

10

Instrumentation

Any basic development in science calls for a new technique fundamentally adapted to the new purpose. [R. J. Van de Graaff, K. T. Compton, and L. C. Van Atta, *Phys. Rev. 43*, 149-157 (1933).]

Progress at the limits of humankind's understanding of the physical universe must always await the development of more-powerful scientific instruments—instruments that enable us to resolve finer features or detect fainter signals. It is a lesson of history that the new information that these instruments provide will eventually be applied to satisfy some need of society. Because the time between discovery and application is often decades, however, basic science is generally treated as a long-term investment and hence tends to be sacrificed in competition with more immediate needs. Often overlooked, however, is the overwhelming tendency for the instruments themselves to find application long before the fundamental science they are developed to reveal. Moreover, these instruments, both large and small, are often applied in ways beneficial to society that are totally unanticipated at the time of their development.

As might be expected, the tools of pure science are first taken up by the more applied branches of science. Nuclear magnetic resonance (NMR), for example, was conceived by physicists as a means of measuring the magnetic moments of atomic nuclei, a quantity of fundamental importance. When placed in an external magnetic field, atoms with a net nuclear dipole moment align themselves either parallel or antiparallel to the field, which has the effect of creating two energy states for the nuclei. A NMR experiment consists of inducing transi-

185

tions between these energy states by superimposing a small oscillating magnetic field upon the strong uniform field.

The observation of NMR in solids and liquids depends on the relaxation of the nuclear spins to the lower-energy state (parallel with the field). This relaxation takes place through the interaction among nuclei in the lattice. NMR studies have therefore yielded a considerable body of information on the condensed-matter properties of solids. This spin-lattice interaction is particularly striking in the case of protons (hydrogen nuclei) in organic compounds. It results in a fine structure in the proton resonance line that provides the basis of an important tool in analytical chemistry. Thus, NMR, which began life as an instrument of basic science, was adapted to progressively more applied sciences.

Quite recently NMR has taken the step from scientific instrument to tool of medical diagnostics, a step that could not have been foreseen by its developers. Whole-body scanning by NMR provides sectional images of the human body of remarkable clarity and with none of the potential hazards of x-ray scanning. Today this is called magnetic resonance imaging. There is now discussion that NMR may one day allow doctors to observe human metabolism without surgical procedure, moving us one step closer to the possibility of knifeless biopsy. Clinical results indicate that NMR is particularly good at detecting dead tissue, lesions such as muscle lesions in multiple-sclerosis victims, malignancies such as pituitary tumors, and degenerative diseases of various kinds. By virtue of this success, it has been projected that the worldwide market for NMR scanners may exceed $1 billion within a few years. The apparatus needed for NMR imaging uses much of the same technology developed by basic researchers for NMR spectroscopy. Many early NMR imaging experiments were, in fact, performed with modified NMR spectrometers. What has made the modern imaging machine possible is the coupling of the spectrometer and the computer. Indeed, the computer has become ubiquitous in modern instrumentation and is now incorporated into virtually every new measurement system. Chapter 13, on medical applications of physics, describes additional instrumentation developments that are advancing medical diagnostics to unprecedented capabilities.

Some remarkable examples of unanticipated applications of scientific instruments have come in the area of particle accelerators. Developed initially to probe the structure of the atomic nucleus and finally to unravel the structure of the subnuclear particles, these machines were conceived as instruments of pure science with little or no foreseeable application.

As accelerators provide answers to the questions that they were designed to address, they become obsolete as tools of basic research. The Van de Graaff accelerators, which were used to accelerate protons to energies of from 100 keV to perhaps 1 MeV, were once the mainstay of nuclear research. Today they have all but disappeared from nuclear research laboratories. But despite their decline in nuclear research, there are today more Van de Graaff accelerators in use than at any time in the past, and the number is still growing. More than a thousand are used in the semiconductor industry alone, where they are used to implant dopant atoms into microcircuits. These foreign atoms are driven into semiconductor elements too small to be seen by the most powerful optical microscope where they precisely tailor the electronic properties of the semiconductor. In the United States alone, annually some $8 billion worth of products that go into computers and microprocessors depend on this implantation procedure. In addition, several companies have developed small, self-contained Van de Graaff generators for use by the petroleum industry as an aid in probing geological strata deep underground in search of gas and oil deposits.

As the Van de Graaff accelerators disappeared from the nuclear research laboratories, their place was taken by a generation of more powerful accelerators, the cyclotrons, capable of accelerating protons or heavier ions to energies up to 200 MeV. These new machines allowed scientists to examine the nucleus of an atom in much greater detail than was possible with the Van de Graaff accelerator. But, again as a measure of progress in the field, the number of cyclotrons used in basic nuclear research has sharply declined. Cyclotrons are, however, still being manufactured for their ability to transmute elements, and today they produce many of the radioactive isotopes used in medical research, diagnosis, and treatment.

One of the unanticipated benefits of electron accelerators has been the synchrotron radiation produced when the path of a charged particle is bent by a magnetic field. The intense synchrotron radiation, stretching from the visible through the x-ray wavelengths, was regarded initially by accelerator designers as a detriment since the radiation carries off energy; it has since made possible experiments in fields such as solid-state physics that are quite unrelated to the fundamental-particle physics for which the synchrotrons were designed. The result is that today major synchrotron facilities are constructed as dedicated light sources for experiments requiring a broad band of frequencies. Specially designed magnets, appropriately called wigglers and undulators, steer the electron along swerving paths to enhance the available synchrotron radiation. Moreover, the intense radiation in the x-ray region may turn out to be the best source for the x-ray lithographic techniques that will be an essential part of the next generation of microelectronics. It is therefore conceivable that storage rings will eventually be incorporated into integrated-circuit manufacturing plants around the world.

Today, Fermi National Accelerator Laboratory (FNAL) in Batavia, Illinois, houses an accelerator that has achieved energies of nearly a trillion electron volts, and a far larger accelerator is now contemplated. While the information that these machines provides enriches our view of the universe, the machines themselves may have uses that are not now foreseen. In any case, former FNAL director Robert Wilson's famous statement before the Joint Committee for Atomic Energy is germane: "[Fermilab] has nothing to do directly with defending our country except to help make it worth defending."

Despite the importance of such major user facilities as Fermilab to scientific progress, however, most of the physics research in the United States is conducted in small university laboratories consisting of a few researchers and a total instrumentation inventory of less than $1 million. Research on this scale continues to produce much of our fundamental new science and to train a large fraction of our scientific talent. It is also on this scale that many of the most significant innovations in instrumentation occur.

Nowhere is this better illustrated than in research on solid surfaces. The problems of preparing and maintaining clean or controlled surfaces do not lend themselves easily to shared facilities. Techniques for the spectroscopic analysis of surfaces have, however, been developed during the past two decades in numbers that strain our capacity to devise new acronyms. Will this period of fecundity continue indefinitely, or is the end imminent as one by one we exhaust all the possibilities? Some idea of the number of possibilities might be gained from a consideration of the diagram in Figure 10.1. The circle represents the sample to be analyzed. Ingoing arrows correspond to the possible surface probes; outgoing arrows correspond to the secondary particles, which convey information about the sample. Each combination of an arrow in and an arrow out would seem to constitute a potential analytical technique. There are, however, only 36 such combinations, all of which have been tried at some time and about half of which are in more-or-less common use. Clearly there exist many more techniques than this. The profusion of spectroscopies stems from the fact that a single combination of an arrow in and an arrow out may lead to several quite dissimilar spectroscopies, depending on what properties of the probe and emitted particle are measured. The change

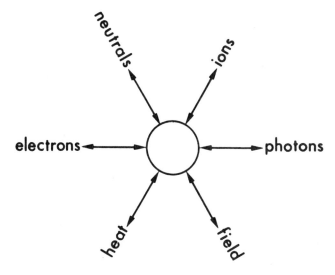

FIGURE 10.1 Diagram illustrating the profusion of surface spectroscopies. The circle represents the sample to be analyzed. Ingoing arrows represent the various probes used to excite the sample. The possible responses to that excitation are indicated by the outgoing arrows, which convey information about the sample. Every spectroscopy can be represented by a combination of an arrow in and an arrow out, but a single combination may lead to several dissimilar spectroscopies, depending on what properties of the probe and the output are measured.

in momentum of an electron elastically scattered from a surface, for example, can provide information on the arrangement of its atoms. The measurement of characteristic energy losses of scattered electrons, on the other hand, has produced information on such diverse phenomena as core-level excitations, collective oscillations of the electron gas, and vibrational modes of surface atoms, depending on what energy range is involved. Thus the number of new possibilities seems unlimited. In addition, older techniques that have been tried and rejected continue to be rediscovered as their limitations are obviated by advances in technology or the interpretation of their output is made possible by new theoretical models.

A unifying feature of the surface spectroscopies is their reliance on a vacuum environment. It is a marvelously unreactive environment. Even at a pressure of 10^{-12} atm, a complete monolayer of contamination can condense on a surface in less than an hour. Such pressures, which are orders of magnitude lower than the pressure outside the Space Shuttle in orbit, were at one time painstakingly achieved in only a handful of laboratories. With modern vacuum technology, however, such pressures are routinely achieved with standard commercial vacuum equipment.

But if surface science has been a beneficiary of modern vacuum technology, it has also been one of the principal contributors to that technology, and indeed, most of the major advances in vacuum technique were pioneered by surface physicists. Today the entire microcircuit fabrication industry relies on these techniques.

Perhaps the most fundamental problem of surface physics is to determine the arrangement of atoms in the outermost layers of a solid. Given this arrangement, modern computational techniques make it possible to calculate many of the most important surface properties. Determination of surface structure is not, however, a problem that has been easily solved. But there are today a variety of techniques, including low-energy electron diffraction, Rutherford backscattering, x-ray standing waves, and angular-resolved photoemission, that contribute to this determination.

Low-energy electron diffraction (LEED), which gives an essentially statistical view of a structure, is particularly useful in the important area of two-dimensional phase transitions. The adsorption of simple gases on single-crystal surfaces frequently takes the form of ordered two-dimensional crystals in registry with the substrate. Until recently the emphasis in LEED studies of adsorption has been to deduce the structural details of these static ordered overlayers. Such structures, however, often exhibit order-disorder transitions, and there has been increasing interest in using LEED to follow these phase changes in the adsorbed layer.

There are two reasons for this increased interest. For one thing, the chemisorbed layers represent a good physical realization of so-called lattice-gas models in two dimensions. Several symmetry classes of such two-dimensional statistical-mechanical models can be solved exactly, whereas this has not been possible thus far in three dimensions. Exactly solved problems are rare, and yet they provide much of our conceptual understanding of nature.

A second reason for interest in the details of the two-dimensional phase diagram of adsorbed layers is that it offers a means of determining the complex interactions among adsorbed atoms. These interactions are fundamental to understanding surface reactions.

Unlike the study of ordered structures, however, which requires only an analysis of diffraction intensities, the study of disorder requires a precise measure of the broadening of diffraction beams. This has led several laboratories in recent years to undertake the development of a whole new class of high-resolution LEED systems.

For many purposes, however, it is desirable actually to image the surface atoms themselves. It is interesting to note that the first microscopic images of surface atoms were not obtained, as one might expect, as a result of a major research effort involving the talents of many people and the commitment of large facilities. Rather they were obtained with an instrument of exquisite simplicity produced by a single researcher working alone in a modest university laboratory. The field ion microscope, developed by Erwin Müller more than 30 years ago, uses the high electric fields that can be created at a sharp metal point to ionize helium atoms, which are then accelerated by the field to a fluorescent screen. The ionization takes place preferentially in the vicinity of a surface atom with the result that in the magnified image of the metal point that appears on the fluorescent screen the individual surface atoms are resolved.

Although field ion microscopy represents one of the most aesthetically satisfying experiments in physics, its use is limited to especially sharp points of a few refractory metals, and efforts have continued to find microscopic techniques that can resolve individual atoms on less special surfaces. Success came from two quite different approaches. One obvious approach was simply a step-by-step improvement in the resolution of the scanning electron microscope. The most advanced product of this approach is a 1-MeV microscope at the National Center for Electron Microscopy at Lawrence Berkeley Laboratory [Figure 10.2(a)]. The microscope incorporates numerous microprocessors and high-speed electronics to stabilize the voltage and to correct instantly for small vibrations and distortions in the apparatus. Under some conditions the microscope is capable of atomic resolutions [Figure 10.2(b)].

It is interesting to note that the scanning electron microscope (SEM) played an essential role in the development of modern microcircuits. Indeed, the elements of the newest random access memory chip are too small to be resolved with the best optical microscopes and can be viewed only with the SEM. Thus the microprocessors that the scanning electron microscope helped to make possible have become an integral part of a new generation of SEMs that may one day contribute to the fabrication of devices on an atomic scale.

Atomic resolution has also been achieved with the scanning tunneling microscope (STM), a device whose genealogy is quite different from that of the SEM. The STM operates by quantum-mechanical tunneling, whereby a particle can spontaneously cross a barrier even

FIGURE 10.2 *Top:* 1-MeV scanning electron microscope at the National Center for Electron Microscopy at Lawrence Berkeley Laboratory. *Bottom:* Image of zirconium dioxide crystal of atomic resolution obtained with a 1-MeV microscope. The microscope was constructed in Japan. (Courtesy of the National Center for Electron Microscopy, Lawrence Berkeley Laboratory.)

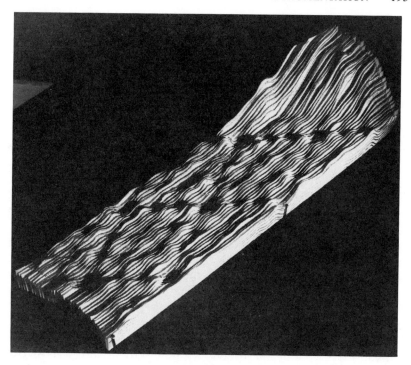

FIGURE 10.3   STM image of a silicon surface assembled from individual contour scans. The single-crystal silicon surface differs from the bulk structure. The image clearly shows two rhomboid-shaped unit cells with atomic resolution. In contrast to the 1-MeV electron microscope, the STM fits on a tabletop. The image was obtained by Bennig, Rohrer, Gerber, and Weibel of IBM Zurich. (Courtesy of the American Physical Society.)

though the particle does not have sufficient energy to surmount the barrier classically. In its operation, tunneling is promoted across a vacuum gap between a sharp metal tip and the sample's surface. A constant-tunneling current is maintained across this gap. This requires holding the tip at a fixed distance above the closest surface structure since the tunnel current depends exponentially on the tip to surface distance. The displacements of the tip given by the voltages applied to piezodrives then yield a map of the surface geometry and composition on an atomic scale, as shown in the image of a silicon surface in Figure 10.3.

The STM has overcome difficult technical problems. Whereas the tunneling current is sensitive to the tip/surface distance, the tip/surface distance has to be kept within about 10 Å (or one billionth of a meter) with a stability of 0.1 Å. To achieve such unprecedented resolution, rigorous vibrational isolation is required. Also, although an ultrahigh vacuum is not required for observation of vacuum tunneling, it is needed to ensure surface cleanliness. A remaining problem whose solution is still eluding researchers is the ultimate control of the sharpness of the metal tunneling tip.

It is nevertheless clear that the successful operation of the STM has opened a new area of surface studies. About two dozen of these microscopes are currently in operation in laboratories around the world.

Each new advance in instrumentation is made possible by the technologies that have resulted from previous generations of scientific instruments. This is quite apparent in low-temperature physics, where researchers strive for ever lower temperatures to uncover new, often unexpected phenomena, such as the startling discovery of the superfluidity of helium-3 in 1971. The lowest temperature achievable has decreased by a factor of 10 roughly every 15 years. Record low-spin lattice temperatures of 50 nK have been achieved in Helsinki, Finland, while in Tokyo the Japanese have attained record-low lattice temperatures of less than 20 μK in metals.

Such unprecedented cooling could not have been accomplished without the development of the dilution refrigerator, a powerful instrument that alone is capable of reducing temperatures to millikelvins. The dilution technique, first proposed in 1962, operates by adiabatic dilution of helium-3 in liquid helium-4. It is currently used in research laboratories worldwide to provide an additional stage of cooling in the study of low-temperature properties of matter. In addition, it has found a direct application in the recent development of an acoustic microscope which uses liquid helium as the operating medium and can resolve features of a few angstroms in size. In astronomy, infrared radiation detectors, which recently provided us with an entirely new view of the universe, require low temperatures for operation. Currently, they operate at 300 mK, and the dilution refrigerator can be expected to provide still better resolution.

The combined use of low temperatures with high magnetic fields, which is critical in the development of superconductors, has only recently led to the discovery of the fractional quantum Hall effect. A two-dimensional conductor undergoes a phase transition at low temperatures and high fields to a state in which the electrical resistance of the substance varies with current in steps of h/e^2. A serendipitous result of this new field of study will be improved measurement standards of electrical resistance and greatly improved determinations of certain fundamental constants.

Superconducting magnets, high-speed electronics, laser interferometry, and countless other technologies that contribute to instrument development all came about originally as unanticipated benefits of earlier instruments. By coupling these new technologies to the computer, it is now possible to consider experiments of a complexity and sophistication that we could not have imagined a few years earlier.

Inevitably, however, as instruments become more sophisticated, they also become more expensive. In 1970, the National Science Board commissioned a study by the National Research Council of the National Academy of Sciences and the National Academy of Engineering on the status of instrumentation in university research laboratories. At that time a need for a $200 million investment in new instrumentation was identified. In the intervening decade, the consumer price index rose by slightly more than a factor of 3, while instrumentation costs inflated at almost twice that rate. The accumulated need is now at least $1 billion, and some estimates are considerably higher.

This estimate is based only on bringing university research laboratories up to modern standards and does not include major new facilities such as reactors and accelerators. More recently, results from a 1984 National Science Foundation instrumentation survey of selected universities indicated that only 16 percent of current academic research equipment in the physical sciences could be characterized as state of the art. The majority of physics research in the United States is carried out in laboratories consisting of a few researchers and a total instrumentation inventory worth less than $1 million. Research on this scale continues to produce most of our fundamental new science and to train a large fraction of our scientific and technical personnel.

There are, however, many important areas that are completely dependent on large centralized research facilities such as the large accelerators. The scale of these projects often makes it increasingly impractical to locate them on university campuses, and many are located at the national laboratories where they are accessible to scientists from all over the United States and other countries. Whereas travel to these centralized research facilities has become an accepted way of life for scientists working in nuclear physics and high-energy particle physics, the need for major centralized research facilities is increasing for work in areas such as condensed-matter physics.

This has been particularly true in the use of synchrotron light sources. The use of photons as a probe of matter is not new, and techniques such as x-ray photoelectron spectroscopy, ultraviolet photoelectron spectroscopy, and extended x-ray absorption fine-structure analysis were all developed using compact photon sources that were accessible to individual researchers. The synchrotron light sources, however, with their high intensities and broad spectral range, make possible whole new classes of experiments that are vital to maintain the lead of the United States in areas of science that are closely linked to our electronics industry.

A relatively new national user facility at Cornell University is devoted to research on the fabrication and properties of structures with dimensions in the submicrometer range. On this scale the wavelength of electrons in the structure may match the dimensions. As a result, the structures can have electronic properties that are determined by feature size. It is on this scale that future generations of random-access memory chips are being fashioned.

In some types of centralized facilities, the United States has clearly failed to keep up with progress in the rest of the world. In the past decade, for example, neutron-scattering research worldwide has shown a rapid expansion both in the number of users and in the diversity of disciplines in science to which neutron methods are applied. This growth is attributable to the development of improved sources and new instruments, which have greatly enhanced the energy and wave-vector range, resolution, and sensitivity of neutron experiments. The United States, however, has fallen far behind Western Europe in the development of advanced facilities at research reactors.

11

Applications of Physics to Energy and Environmental Preservation

ENERGY

The connection between energy and physics runs deep. The concept of energy is a cornerstone of our science; the principle of conservation of energy is one of the few pieces of solid ground in the shifting sands of theoretical physics. In its popular meaning the word energy denotes power, the use of machinery to relieve humanity from brutal labor, the steam engine, the electric motor, modern transportation, heat and light; all these are fruits of physics of the nineteenth and twentieth centuries. Thermodynamics and electricity are keystones in energy production, and they are simultaneously names of courses in the physics curriculum. The question, then, is not so much how physics is related to energy but what have been the contributions of physics research over the past 10 or 15 years to what is commonly described as the energy problem.

An examination of the energy situation reveals that both the euphoria of the 1960s and the panic of the 1970s seem to have dissipated.

Although the dramatic increase in the cost of fossil fuel has caused economic dislocation, it has not resulted in a revision of the world's energy supply. The crisis of the past decade has illustrated that increased price of energy leads to lower demand and much greater emphasis on conservation. For the next 30 years, and barring some truly disruptive political developments, there seems to be a reasonable

197

supply of both oil and gas at slowly increasing prices. However, the cost of nuclear fission power plants has escalated so rapidly that development of this new technology now appears further in the future than many had expected.

Energy matters are, however, not static. For example, the higher price of oil and gas has led to increasing research in combustion and in high-temperature materials to enhance thermal efficiencies of engines. Solar energy does not seem likely to replace central-station electric installations; but it does have an increasing role to play in special power sources and low-demand isolated power stations. Fission, fusion, and solar are all three energy sources inherently capable of providing humankind with continuous energy supplies once the economically available fossil fuel is exhausted. Of these, fission energy is technologically and commercially in use, solar energy awaits technological and cost breakthroughs to serve as a replacement on a large scale, and fusion has yet to prove its utility as a power source. We must recall that the world has vast resources of hydrocarbons in the form of shales and tar sands, and perhaps other yet undiscovered geologic forms, in addition to coal. These will compete for at least several hundred years with fission, fusion, and solar technologies.

Energy production is an enormous enterprise. In end-use cost it amounts to about 12-15 percent of the U.S. gross domestic product. Because of past economies of scale much of our energy economy is highly centralized; therefore its effect on the environment is significant. Finally, we note that our ability to use fossil fuels may in the long term be limited not by the supply but rather by the increasing concentration of carbon dioxide in the atmosphere, which could have truly catastrophic consequences.

The relationship between physics and energy production and use does not always follow the simplistic course that physicists' discoveries are later adapted by engineers for practical use. Indeed it sometimes works out this way, but just as frequently the requirements of energy technologies motivate physics explorations. Frequently advances in technology thus make certain kinds of physics research possible and thus move the science forward.

Condensed-Matter Physics and Solar-Energy Conversion

The transformation of sunlight into electricity, heat, and fuels can be realized by a variety of technologies. This variety arises because the solar radiation may interact directly with the conversion device, as in photovoltaic, photothermal, and photochemical systems, or it may first interact with the biosphere to permit more indirect energy generation, such as wind and water utilization, ocean thermal gradients, and biomass conversion. The wide range of conversion modes gives rise to a multiplicity of materials-related problems. Because photovoltaic conversion is the solar technology that relies most heavily on advanced research in condensed-matter physics, we have chosen to discuss briefly the two recent illustrations of the intimate interplay between basic research and advanced photovoltaic technology.

Today the highest efficiencies measured of single-crystal silicon solar cells designed for terrestrial operation under one-sun conditions are slightly over 19 percent. The efficiency record will soon be pushed to over 20 percent. For comparison, a plant converts sunlight into biomass at about 1 percent efficiency. This remarkable improvement came about quickly, primarily because the solid-state physics involved in these devices has been clarified by theoretical considerations and experimentation. More specifically, it was realized that every mechanism leading to recombination of the electrons and holes generated by the absorption of solar photons must be suppressed. This realization led to methods for passivating the surface by eliminating surface recombination sites introduced by dangling bonds, defects, and processing-induced imperfections. It was necessary to passivate not only that part of the surface exposed directly to the sunlight but also the area covered by the thin metallized fingers used to collect the photogenerated current. Passivation of the metallized area required the growth of an oxide layer that was thick enough to prevent direct contact of the metal with the silicon surface yet thin enough to allow quantum mechanical tunneling of the charge carriers through the oxide. It is no exaggeration to assert that these advances could not have been made without the groundwork laid by advanced solid-state research during the past decade.

A second dramatic advance in solar technology is the recent improvement of the efficiency of amorphous silicon (a-Si) solar cells to just above 10 percent. This improvement was brought about primarily by advances in the basic understanding of the structure of a-Si and the role of hydrogen in reducing the density of localized energy states as well as through a greater appreciation of the role played by residual impurities from the gases used for depositing the silicon. Intensive fundamental research studies of the properties of a-Si alloyed with large concentrations of hydrogen and other elements have led to methods for varying the band gap and controlling the density of gap states. This kind of research can be expected to yield further advances in the efficiencies of a-Si solar cells. Moreover, with the established capability of varying the band gap, the possibility of fabricating stacked or multijunction cells arises. It has already been demonstrated that two-layer *p-i-n* cell structures can be processed. In such a cell, each *p-i-n* layer, although it is only 0.5 μm thick, is doped to form thin *p*- and *n*-type regions separated by a thicker insulating region, and as a consequence each cell may be composed of as many as six different deposited layers with widely different electrical properties. Controlling both the bulk properties and the interfaces between these layers is of the utmost importance. The possibility of fabricating multijunction a-Si cells with efficiencies as high as 30 percent emerges as our basic understanding of amorphous solids and how to prepare them in thin-film form advances.

One common thread linking many types of solar-energy conversion systems is the crucial role played by the interfaces between various types of materials and between differently oriented grains of the same material. Research on the properties of surfaces and interfaces is one of the most active, and still rapidly growing, areas of condensed-matter physics. This research area has received much of its impetus from the requirements of energy-related technologies.

A third feature of solar-related research is the use of electron, ion, and laser beams in materials processing. Pulsed-laser processing for the fabrication of solar cells and other semiconductor devices stimulated intense interest in, and conflicting interpretations of, the effects of ultrashort (10^{-8}- to 10^{-14}-s) laser pulses on semiconductors. While some contended that the near-surface region melted, others ascribed the observed effects to a high-density electron-hole plasma. The controversy has now been resolved in favor of the melting theory. It stimulated considerable research into the interaction of intense laser pulses with solids. Solidification velocities of 5-15 m/s are readily attainable during pulsed-laser melting and resolidification. Reaching such high velocities under well-controlled conditions has greatly increased our understanding of solidification processes occurring far from thermodynamic equilibrium. The same nonequilibrium processes have made it possible to achieve high-efficiency solar cells by tailoring dopant profiles in the near-surface region to passivate the surface effectively through the action of strong built-in electric fields, thus eliminating the need to grow oxide layers. These areas of research will undoubtedly continue to be actively pursued for some time. This is surely one of the most dramatic recent illustrations of the synergistic interplay between fundamental and applied research that frequently occurs in physics.

These advances in understanding and in solar-electric conversion efficiencies for both single-crystal and amorphous cells are necessary steps toward devising a practical technology. Progress in reducing the cost of these devices now makes solar electric conversion competitive in on-site, low-power applications.

It must be borne in mind that solar-energy generators cannot supply energy at a constant rate. They need storage devices. In none of these areas, except for high-temperature batteries, has modern physics yet played a significant role. Opportunities abound.

Finally, one may point to the pressing need for advanced, nondestructive, analytical techniques to probe the effects of various processing steps on the electrical properties of devices without the need to complete the fabrication of the device or system. Techniques of nanosecond, picosecond, and femtosecond optical spectroscopy now being developed for fundamental studies may eventually be used to provide in situ monitoring of the device-fabrication sequence. Other time-resolved measurements, such as electrical conductivity, on the nanosecond and subnanosecond time scale may also be developed for similar and other advanced technological applications.

Materials

Condensed-matter physics plays a critical role both in our understanding of the properties of materials and in our use of them for specific energy-related application. The most familiar of these materials is silicon, but perhaps closer to our discussion of energy is the physics of superconductors.

High-temperature superconductors are currently used to obtain high magnetic fields, potentially up to 10 T, and to conserve energy in large accelerators. They are likely to be essential as well for the eventual harnessing of controlled fusion.

Most fusion-reactor designs that rely on magnetically confined plasma employ superconducting magnetics. At present NbTi, which is limited to use in magnetic fields of less than 10 T, is the most widely used material. Higher-field materials such as Nb_3Sn can be used, but their brittleness makes fabrication difficult. Improved understanding of the basic processes that determine the material characteristics of superconductors should permit the development of new materials as well as the improvement of existing ones. Cheaper and higher T_c materials could show economic advantages in superconducting transmission lines, superconducting windings in large motors and generators, and perhaps in linear induction motors for rail transport.

Other less exotic materials made possible by recent advances in physics are also important in conserving energy. Metallurgists and ceramists now use tools of modern physics such as Auger spectroscopy; neutrons and x rays; electrostatic particle accelerators; lasers to produce new alloys that may be stronger, more corrosion or radiation resistant, cheaper, or more ductile; and ceramics that have greater toughness or improved surface properties. Each of these contributes to energy conservation.

Fusion Energy

The conversion of energy to useful forms is an age-old activity. Two modern processes are nuclear fission and fusion. Both are outgrowths of twentieth-century physics. We do not examine fission energy here because so much has been written about it elsewhere. Magnetic-confinement fusion, on the other hand, is a new and developing technology, whose eventual accomplishment depends crucially on physics. Specific research areas include (1) the atomic physics of multiple-charged ions in a high-temperature plasma (including radiation, ionization, and recombination-charge exchange); (2) interaction of energetic ions, neutrals, and electrons with complex surfaces (including sputtering, reflection, and desorption); (3) the development of plasma diagnostic techniques; and (4) the underlying physics required for fusion power (radiation damage, superconductivity, coherent radiation sources, and particle-acceleration physics). The specific roles of plasma physics, diagnostic development, and atomic physics are discussed in the volume on the Physics of Plasmas and Fluids. We touch on the other areas in the following paragraphs.

PLASMA-SURFACE INTERACTIONS

As a result of both plasma and atomic processes, the surface of the confinement vessel that faces an energetic plasma, the first wall, is bombarded by protons, electrons, ions, and neutrals of widely varying energies. The surface responds in a variety of ways, including elastic and inelastic reflection, absorption, and emission of atoms on the surface (hydrogen, pure wall constituents, and impurities), to name a few. The particles that are emitted or reflected back into the plasma can substantially modify the behavior of the plasma. The details of the wall response depend on both the intrinsic character and the history of the surface. Microscopic models of these processes are required in order to formulate appropriate boundary conditions for comprehensive plasma models. In particular, one deals here with differential sputtering cross sections, releases of occluded gases, and other aspects of surface physics and surface interactions. More detailed knowledge would permit the design of first-wall materials and structure, which would lead to better plasma performance.

RADIATION EFFECTS IN FUSION REACTORS

The first-wall and structural components of a fusion reactor will be subject to high fluxes of high-energy neutrons released in the reaction $D + T \rightarrow He + n + 17.6$ MeV. Of this energy, approximately 14.1 MeV is the kinetic energy of the neutron. These highly penetrating neutrons are slowed down in the first wall, blanket, and structure, inducing severe degradation of mechanical properties and dimensional instability in the structural materials of the reactor.

Energetic neutrons displace atoms in the alloy crystal lattice, producing vacant lattice sites (vacancies) and atoms in interstitial positions (interstitials), collectively known as point defects. Neutron irradiation also induces transmutation reactions of the alloying constituents. The transmutation products take the form of both solid and gaseous impurities. The most important of these products is helium, a highly insoluble rare gas, which will reach concentrations of a few tenths of atomic percent in a fusion power reactor. These radiation-induced defects are ultimately responsible for the progressive deterioration of reactor materials with increasing dose. Most vacancies and interstitials recombine with one another, leaving permanent damage. However, a small fraction agglomerate into clusters. Interstitials condense into platelets called dislocation loops. Vacancies cluster as voids, a new phase of empty space within the crystal. The point defects also diffuse to pre-existing sinks, such as dislocations and grain boundaries. Helium diffuses to grain boundaries. The flow of point defects to sinks induces fluxes of alloying elements through diffusional reactions. These fluxes can cause progressive demixing of the alloy, leading to highly segregated (enriched or depleted) regions on the scale of nanometers and to harmful radiation-affected precipitation reactions.

Drastic changes in properties can occur through these reactions. Embrittlement, loss of ductility, and cracking can be caused by the presence of helium on grain boundaries, segregation of alloying elements and impurities to grain boundaries, and hardening of the grains by point-defect clusters and precipitates. For example, in high-dose neutron irradiations in fission reactors, structural materials have been reduced from tensile ductilities (elongation without failure) of tens of percent to tenths of a percent. In materials not designed to be resistant, swelling can reach values of 50 percent or more. Irradiation creep, a permanent volume-conserving deformation in the direction of an applied stress, can be orders of magnitude greater than the ordinary thermal creep of structural materials observed at elevated temperatures. It results from stress-induced nonsymmetrical flows of vacancies and interstitials to dislocations.

Research using fission reactors or charge-particle accelerators has revealed much about radiation effects under fusion reactor conditions. By using these tools, much better alloys are being developed by tailoring microstructures and compositions. For example, improved stainless-steel alloys swell as little as 1 percent under the conditions in which unimproved alloys may swell many tens of percent.

COHERENT RADIATION SOURCES AND PARTICLE ACCELERATION PHYSICS

A central problem in fusion plasma physics has been the development of efficient systems for plasma heating. Electromagnetic radiation in a wide frequency range and particle beams have been two useful approaches. In many cases the ability to advance the understanding of plasma physics has been limited by available technology. Advances in producing neutral beams and high-power sources of microwaves have led to major improvements in tokamak and mirror fusion devices. Pulsed electron and ion sources have had a similar role in other fusion devices. At present there are well-defined needs for power sources to heat planned fusion experiments. These systems are under development by the fusion program discussed in the volume on the Physics of Plasmas and Fluids. However, there is also the possibility that fundamentally new discoveries of how to produce, for example, tens of megawatts of steady-state 1-mm microwave radiation might in turn generate new confinement configurations in themselves. Similarly, new concepts of how to produce multiampere beams of negative hydrogen ions efficiently would lead to re-evaluation of how best to heat a large fusion device.

INERTIAL-CONFINEMENT FUSION

A totally different way to achieve fusion of deuterium and tritium nuclei is to confine the plasma by inertial instead of magnetic forces. The principle is as follows: take a small hollow sphere, normally less than a millimeter in diameter, fill it with tritium or deuterium along with other inert materials, and then push on it from all sides until the density and temperature are high enough to produce a tiny thermonuclear reaction. The implosive push is produced by the ablation of the surface, and the high power required to ablate the surface at the appropriate rate can be provided by lasers or by energetic heavy ions or other particles. (Much of this technology derives from weapons design and remains in part classified.) The physics involved has to do with (a) the behavior of the sphere, its surface, and its interior at high power densities and (b) the creation of high-powered lasers or accelerators that can focus energy on the sphere.

It is evident that laser- and particle-beam physics plays a key role in this energy-conversion technology, and it is inevitable that our basic understanding of high-powered lasers and nonlinear optics will increase as a result of this applied program. Otherwise, inertial fusion is not so close to becoming a useful source of energy as is its magnetically confined counterpart.

The Role of Physics in Combustion Research

Combustion has always been and remains today responsible for over 90 percent of all energy conversion in the world, yet the physical basis for understanding, predicting, and controlling combustion is only now being given significant scientific attention. The benefits of improved combustion technology are great in both energy gain and environmental control. A 1 percent increase in automotive engine fuel economy can save the United States alone nearly $1 billion a year; combustion control of sulfur in furnaces may ameliorate or at least clarify the acid rain issue; control of combustion-driven corrosion and erosion in boilers can save the utilities billions of dollars in repairs and down time. These issues, and many of the benefits, were highlighted in the early 1970s through a series of studies by several government agencies and professional societies. In 1975 a report of the American Institute of Physics (*The Role of Physics in Combustion*, AIP Conf. Proc. 25), highlighted several areas of physics involvement that could accelerate combustion-science research. The recommendation for physics input focused on research tools—laser diagnostics and computational modeling—as the main adjunct to traditional combustion science and on innovation in turbulent reacting flows or catalytic combustion and ultralean combustion.

In the decade since that report noteworthy advances have been made; however, a great deal of work remains to be done. As old problems are resolved and the technology advances, even more complicated issues are born; this is particularly true in the areas of coal utilization and combustion control in engines.

Opportunities for physics to contribute to combustion science lie primarily in the areas of diagnostics and high-temperature fluid physics. However, many physics concepts contribute to the various aspects of combustion, as described below.

DIAGNOSTICS FOR COMBUSTION

In providing the proper light sources, optics, and detectors for combustion diagnostics, physics is already contributing importantly. For the supporting spectroscopy, requirements to define photon-gas interactions, line shapes, scattering cross sections, and ionization cross sections place renewed emphasis on the study of quantum physics, quantum chemistry, and collisional scattering analysis.

FUEL PREPARATION AND MIXING

Proper definition of the initial phases of droplet formation and evaporation, diffusion and turbulent mixing, char burnout, and char analysis will rely on thermodynamics, statistics, molecular and turbulent transport, and surface physics.

IGNITION

A description of ignition processes, from point sources of spark ignition to controlled homogeneous ignition, must build on plasma physics, ionization, and electron and ion scattering.

FLAME PROPAGATION AND EXTINCTION

The principal challenge of predicting a combustion process requires adequate understanding of the physical and chemical details of flame propagation and flame extinction in the bulk and near surfaces. Fluid physics, thermodynamics, energy and radiation transport, molecular dynamics, turbulent transport, catalysis, and surface physics all contribute to this understanding.

POLLUTANT FORMATION

A primary measure of the acceptability of a combustion process is the magnitude of pollutant formation and the degree to which the pollutant can be controlled in the combustion process. The prediction of mechanisms of formation of these minute species requires better understanding of nonequilibrium reactive scattering, turbulent transport, and multiphase flow. Predicting the formation, and destruction, of particulates requires better understanding of heterogeneous nucleation, flow-through porous media, surface physics, radiation, and statistical mechanics. The cleaning of solid fuels by physical and chemical-physical mechanisms, the incorporation of pollutants during burning in fluidized beds, and the catalyzed and reactive removal of gases and particulates in scrubbers also present challenges to the physicist.

Future Developments

Many areas of physics that are now vigorously investigated have the potential to contribute in the future to the efficient production and benign use of energy. Perhaps the most obvious of these is the field of surface physics. As our understanding of metallic and nonmetallic surfaces grows we can foresee the day when catalysis will become a predictive science, and we shall be able to tailor surfaces to catalyze specific reactions, perhaps even without expensive and scarce materials such as the platinum-group metals or cobalt. There may even emerge from the study of surfaces new ways to combat corrosion. Few areas of condensed-matter physics are being studied more intensively than surfaces. Giant synchrotron light sources, ion-implantation devices, low-energy electron diffraction (LEED), Auger spectroscopy, ion and atom scattering, and radiation of all wavelengths are being used for the study of surfaces. It is not unrealistic to expect that this unprecedented marshaling of resources and people will lead to results that will substantially affect energy-related technologies.

Biophysics represents another frontier whose exploration promises to be expanded by energy applications. The study of an organism's biological structures (for example, membranes and giant molecules) by physical means such as nuclear magnetic resonance and neutron and x-ray scattering should yield a qualitatively enhanced understanding of bioprocesses related to the organism's use and transformation of energy.

The modern apparatus of physics, accelerators, and various precision spectroscopies coupled with vastly expanded computing power should have applications, for example, in polymers, ceramics, chemical processes, metallurgy, and geophysics. All these have the potential to influence the conversion and use of energy in unexpected ways. The field is wide open; opportunities for exploiting the results of physics research are brighter than ever.

ENVIRONMENT

Natural environments consist of major interacting units—atmosphere, hydrosphere, land, and biota. These units are, of course, highly complex systems of diverse and interrelated physical, chemical, and biological factors in themselves. In the broadest sense, environmental science thus encompasses many distinct but highly related disciplines, such as hydrology, ecology, meteorology, and geology. Each of these disciplines is rooted in application of physical principles to some degree. Some authors recognize environmental physics as a discipline in its own right. Even many physicists, however, may be unaware just how numerous and essential are the methods, instruments, and applications of physics to environmental studies. Many, if not most, of these applications are concerned with determining the mechanisms, rates, and pathways for movement of matter (such as air masses, nutrients, radionuclide or pollutant contaminants, and even organisms) and energy through environmental systems. Physics contributes both fundamental principles used to describe environmental processes and a diverse array of methods and equipment for measurement and detection. The following discussions will focus on selected examples of environmental applications.

Atmospheric Science

Understanding of the movement of air masses and associated weather patterns; of the effect of the atmosphere on solar insolation, thermal fluxes, and the Earth's radiation balance; and of the transport and diffusion of particles and pollutants are all derived directly from the principles of mass, energy, and momentum transfer. Most of the meteorological problems associated with environmental analyses of power generation, chemical manufacture, and the nuclear-fuel cycle are, for example, focused on mathematical-physical modeling of phenomena such as diffusion in turbulent fluid flows. Although short-range transport modeling has been successful for predicting air-quality effects on a local scale, the current challenge in atmospheric science is to develop an understanding of the more complex, multiple source, long-range transport problems with regional or global implications.

Acid Rain

One of the current major environmental disputes concerns the determination of source areas for acid aerosols and precipitation in the

United States. The areas thought to be most strongly affected (for example, the Adirondacks and New England) receive pollution originating from multiple distant source areas. No transport model available can, as yet, reliably apportion the constituent pollutants to specific source areas. The ability to identify distant sources of pollution correctly could greatly improve the efficacy of environmental controls in reducing acid precipitation, while curtailing the wrong sources could be an extremely costly error.

Promising new empirical approaches are, however, being developed. It has been demonstrated, for example, that although no single-element concentration can be used to trace regional pollution aerosols to unique sources, ratios of elements in aerosol mixtures can serve as effective regional signatures. A seven-element tracer system has been designed that shows that regional pollution aerosols can be followed into remote areas up to several thousand kilometers downwind. The success of this technique is dependent on an ability to measure concentrations of the seven elements in air samples accurately. For this, neutron activation, an extremely sensitive analytic probe, is used. Samples are placed in a neutron flux, and neutron capture then produces radioisotopes whose characteristic decay patterns identify the elements present.

Another application that may become important in developing regional tracers is the use of stable isotopes not naturally found in significant concentrations in air. Investigations in particle nucleation and growth, radiation-induced reactions in particle-gas ambients, and the catalytic action of surfaces can aid greatly in characterizing the complex morphology and molecular structure of these contaminants.

The effects of increasing concentrations of atmospheric particles is important in evaluating the thermal and radiative balances that will determine long-term climate changes. Atmospheric particles reflect solar radiation and in so doing reduce the incident solar intensity. Satellite measurements of the Earth's infrared and visible radiation can be used to search out integrated as well as local emissivity changes.

It is worth pointing out in connection with atmospheric transport that a fundamental understanding of turbulent flow would greatly enhance our ability to predict movement of air masses. In fact, our inability to deal with turbulent fluids is a significant impediment not only in atmospheric science but also in other areas of hydrodynamics and aerodynamics. Many areas of energy production and use as well as in environmental protection would benefit greatly from a basic understanding of turbulent flow and an ability to calculate from first principles.

Carbon Dioxide Concentration and the Greenhouse Effect

Similarly, the radiation balance of the Earth has long been modeled on the basis of physical law. The effects of human activity on local climate, such as in urban areas, are fairly well understood. On a long-term or global scale, however, the effects of changes in the composition of the Earth's atmosphere from combustion of fossil fuels and addition of pollutants are not readily understood. The interactions of air pollutants with Earth, atmosphere, and global climate are complex, and climatic change resulting from human activity is difficult to distinguish from large natural variations.

The increase in the atmospheric concentration of carbon dioxide is undoubtedly the single most important environmental factor in the production and use of energy, as illustrated by Figure 11.1. One hundred years ago the concentration of carbon dioxide was about 280 parts per million (by volume); it is now 340 parts per million, and the concern is that in the next 100 years this concentration may double. Two fundamental kinds of question arise from this situation. Among the first kind are these: What is the flow and balance of carbon dioxide in the biosphere? What is the rate of absorption of carbon dioxide in the oceans? What are some of the other sources and sinks of carbon dioxide? What are the effects of other ways of generating atmospheric carbon dioxide besides burning fossil fuels?

The second kind of question has to do with the effects of, say, doubling carbon dioxide concentrations over the next 100 years. What will be the effects on the climate, and, in turn, what will the climate effects be on agriculture, the world distribution of deserts and forests, or the heating and eventual melting of the solar ice caps with the resulting increase in sea level of perhaps 5 to 8 m?

Both kinds of question deal with systems that are extremely complex and poorly understood, and, because such large masses are involved, the observable changes occur only slowly in time. The carbon dioxide balance problems are mostly questions for ecology and chemistry, although the mixing problem in oceans clearly falls into the sphere of geophysics. This in fact may be the single most important factor in elucidating the question of sources and sinks for carbon dioxide. The questions dealing with climate are full of physics problems. There are questions of atmospheric physics, transmission of radiation through complex atmospheres, cloud physics, general circulation models of the atmosphere, and generally the complex physics of atmospheric balances including transport of water vapor. Because this problem is so important to the future of humankind, we must endeavor to bring great

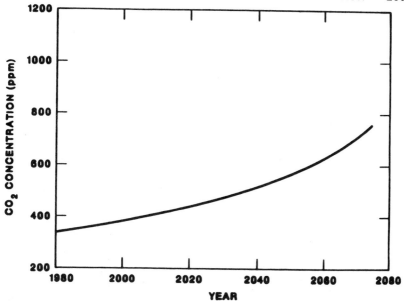

FIGURE 11.1 The projection of the carbon dioxide concentration over the next century depends strongly on the rate of consumption of fossil fuels. Assuming the growth rate to be between an extremely rapid increase in consumption and replacement of fossil fuels by other energy sources, it is likely that the carbon dioxide concentration will increase as shown here.

intellectual strength to bear on the question of control of atmospheric carbon dioxide

Knowledge of the values of critical atmospheric parameters, for example, the concentration of atmospheric carbon dioxide before the large-scale additions from fossil-fuel combustion, is also scant. Such physics-based techniques as isotopic analysis of tree rings and glacial ice are currently being employed to collect evidence to deduce this value. Determination of such factors as the amount of carbon retained in the atmosphere as opposed to what is absorbed by the oceans or vegetation will depend on refinements of the biological, chemical, and physical nature of transport and absorption processes on a global scale.

The problem of the increasing atmospheric concentration of carbon dioxide resulting from the use of fossil fuels is highly important, extremely complex, but fortunately not yet at the crisis level. It is imperative that solution of this problem be assigned high priority and that the research reflect its urgency. We therefore recommend that the federal government support sustained basic investigations in areas such as carbon balance, the absorption in the oceans, radiation transport in the atmosphere, climatic effects, and other areas outlined in this report at a level commensurate with the importance of the consequences of this phenomenon. We recommend that these investigations be organized in a unified program with high scientific visibility.

Hydrospheric Studies

The Earth's water circulates constantly through the various major compartments of the environment by means of the physical processes of the hydrologic cycle—evaporation, precipitation, percolation, and runoff. As in the atmospheric sciences, most environmental analyses that deal with hydrology focus on flow, transport, and dispersion processes.

Groundwater hydrology is, in particular, strongly tied to geophysics for a large body of methodology. For example, such techniques as resistivity depth soundings, seismic refraction depth profiling, and gravity depth modeling are used effectively to identify plumes of contaminated groundwater from abandoned and active waste-disposal sites hidden beneath uncontaminated and revegetated cover soils. Geophysical investigation can, in general, reduce the need for costly test drilling and maximize the amount of information gained when subsurface drilling is unavoidable. Remote-sensing techniques are useful for regional-scale detection problems, and magnetic and thermal geophysical methods are used in special geologic situations. Perhaps one day groundwater hydrology may even be able to take advantage of the new measuring devices being developed with superconducting magnet technology. The superconducting quantum interference device (SQUID), for example, can measure the extremely weak electromagnetic waves that bombard the Earth from space and are reflected by the Earth's surface. Details of the reflection depend on the properties of subsurface materials, and waves can now be used to provide information on subsurface deposits of oil or hot water.

Surface-water analyses pose different kinds of problems because surface waters provide habitats for aquatic organisms and communities. To analyze the effects of a wastewater discharge on a surface-water body, physical principles for transport and diffusion of pollutants must be used in conjunction with biological data on the types and abundances of organisms present and the effects of pollutants on organisms and food chains. The effects of waterborne pollutants on aquatic environments are, however, determined to a large extent by the physical characteristics of water bodies in that the degree of exposure of aquatic organisms will vary with different physical settings. For example, fish inhabiting lakes were found to accumulate higher concentrations of fallout-derived cesium-137, a nutrient analog of potassium, than fish inhabiting streams. Data for trout gathered from Colorado streams and lakes suggested twofold to tenfold higher values for lake fish owing to greater fallout removal processes in streams.

Soil Physics

The field of applied soil physics is of great importance to both agricultural and environmental science. The practical result of soil-physics investigation in general is an ability to modify the microenvironment of plants to create better conditions for growth and higher productivity. For example, measurements of soil matrix potential are used to determine appropriate times to irrigate so that water flow to plant roots does not become a limiting factor in plant growth. Similarly, soil-plant energy budgets can be modeled to estimate rates of evaporation from vegetation surfaces and to calculate water needs of a crop based on climatological measurements. While the rates of water loss and carbon dioxide absorption by crop plants have received much attention in environmental physics, other areas of investigation have been relatively neglected. These include study of factors that determine the temperature, humidity, and wind regimes of crop plants and study of how the microclimates of individual leaves determine plant susceptibility to insects and pathogens.

Biota and Ecology

Although the study of living organisms is the province of the various branches of biology, analyses of biological process and adaptation are frequently performed from the physical point of view. The sources of energy flow between an animal and its environment (metabolism, moisture loss or gain, radiation, and convection and conduction) are usually interpreted in terms of thermodynamic equilibria. Application of the principles of mechanics makes it possible to analyze structural adaptations of and constraints on plant and animal form and function. The abilities of animals to orient and navigate may be understood in terms of responses to physical stimuli (sound and echo location, for example). Thus, at the whole-organism level, biology is substantially tied to physics.

Ecology is concerned primarily with the higher levels of biological organization—populations, communities, and ecosystems—and with the interaction of organisms with one another and with the abiotic environment. The composition of ecological communities is heavily influenced by physical environmental factors, and ecosystem function is driven by energy flow and biogeochemical cycling of nutrients and minerals. Probably the best example of an outgrowth of physics that has had a major impact on ecological studies is the interdisciplinary field of radioecology. Radioecology encompasses the relationships between ionizing radiation or radioactive substances and ecosystem components. (Radiobiology, in contrast, includes radiation relationships at the lower levels of biological organization, such as molecular and cell effects.) Three primary areas of study within radioecology are radionuclide movement within ecological systems and accumulation within ecosystem components; radiation effects on species, populations, communities, and ecosystems; and use of radionuclides and radiation to determine structure and function of ecosystems and their component subsystems. Radioecology is an important interdisciplinary approach to ecology and draws heavily on the basic physical, as well as biological, sciences. Much of the methodology developed in radioecology since the 1940s is used today to study environmental impacts of nonnuclear as well as nuclear energy. Radionuclide tracer studies in general have revealed a great deal of basic information on ecological and biogeochemical processes and can be used as effective means to determine environmental levels of pollutants such as heavy metals.

RECOMMENDATIONS

In order to preserve the United States' energy options and to supply the nation with technology for the production and prudent use of energy, the federal government should support a broad program of basic and applied physics in areas that relate to current and potential energy technologies. Methods for the production of energy that have minimal environmental impact should also be identified and encouraged.

A large part of physics research supported by the U.S. Department of Energy is carried out at this country's national laboratories. We recommend that this practice be continued with special emphasis on the following four criteria: that the research be multidisciplinary or require large groups of investigators; that the research be connected with large facilities or collections of unique equipment that are freely available to outside users from universities, industry, or other laboratories; that the research programs be colocated at a laboratory with closely related technology development programs and thus underpin those programs; and that research at national laboratories be carried out in concert with university and industrial efforts whenever possible.

We recommend strong support for a long-range research program to gain fundamental understanding of all factors influencing the Earth's thermal energy balance, such as changes in the emissivity of the atmosphere and in particular the carbon dioxide concentration and its association with fossil-fuel burning.

12

Physics and National Security

INTRODUCTION

Among scientific disciplines, physics is without peer as a source of discoveries that serve as the basis for national-security strategies and tactics. The profound effects of physics are apparent both in the development of weapons systems and in the complex processes of arms control. Although the contributions of physics are perhaps most notably unique at the basic conceptual stage, physicists typically contribute at any or all of the stages from generation of the basic elements of knowledge and understanding necessary for invention to take place, to invention itself, to the evolutionary improvements and development necessary for successful operational use and exploitation. To all such stages of activity, physicists bring a perspective that focuses on novelty and that demands attainment of in-depth understanding expressed in mathematically rigorous terms. In this century until the early 1970s this perspective was manifest in essential contributions by physics to such revolutionary national-security-related advances as long-distance radio communications, vacuum-tube electronics, underwater sound surveillance, radio navigation and direction finding, radar, proximity fuses, nuclear weapons and schemes for detection of nuclear explosives, nuclear propulsion, automatic guidance and control systems, rockets, electronic computers, infrared

216

night-vision systems, transistor electronics, guided-missile technology, integrated microelectronics, laser technology, internal-computer-controlled "smart" weapons, and satellite-based communications, surveillance, and meteorological systems.

During peacetime, the great majority of U.S. physicists engage either in expanding the store of knowledge in physics or in applying physics to problems in the civil and commercial sectors. In times of national emergency, however, physicists, in large numbers, have refocused their efforts in support of national-security interests. During World War I physicists made crucial contributions to underwater acoustic techniques for hunting submarines and to seismic and optical techniques for locating hostile artillery. During World War II physicists were mobilized on an unprecedented scale. They greatly refined underwater sound techniques for locating submarines, and they made critical contributions to the new discipline of operations research, which had a revolutionary effect on the optimum utilization of military resources. Radar was transformed into a highly sophisticated and extremely effective means of surveillance through a large collaborative effort by physicists and electrical engineers. The associated development of microwave techniques paved the way for invention of the maser (the forerunner of the laser), for the study of electron spin resonance, and for understanding the nature of superconductivity.

By far the most ambitious wartime mobilization ever of scientists and engineers took place in the Manhattan Project. Conceived and intellectually driven by physicists, this effort gave birth to the nuclear age and has profoundly altered the course of history. Again physics was itself altered as new realms for exploration emerged; for example: neutron-scattering determinations of crystal and magnetic structures as well as phonon and other excitation spectra; nuclear radiation effects in materials as they relate to nuclear power reactor engineering; and stable and unstable isotopes, their interactions, and their applications to chemical analysis and nuclear medicine.

EXAMPLES OF RECENT CONTRIBUTIONS OF PHYSICS TO NATIONAL SECURITY

Since publication of the previous physics survey (*Physics in Perspective*, National Academy of Sciences, Washington, D.C., 1972), advances in national-security-related physics have continued. Those presented below are illustrative of a much larger set. Although these examples are largely developmental in nature, they are firmly founded on basic advances in physics of both recent and earlier vintage. Their development has required the collective efforts not only of physicists but of many other scientific and engineering disciplines as well.

Lasers and Their Applications

National-security-related applications of lasers are so diverse and widespread that it is possible only to provide a few examples here. Physicists have made essential contributions to the invention and/or development of a variety of lasers, for example, solid state, liquid, dye, gas discharge, gas dynamic, chemical, optically pumped, injection, excimer, and free electron.

An ambitious goal is the development of laser systems of such high power that the interaction of the laser radiation alone is capable of defeating the target by inflicting damage either on vulnerable components of its nervous system (sensors, radomes, electronics) or on structural components. Although the latter goal remains far in the future, there has been remarkable progress in this direction over the past several years. In 1983 a high-power gas-dynamic carbon dioxide laser successfully demonstrated, in flight, the negation of air-to-air missiles and of drone aircraft simulating antiship cruise missiles. The associated pointing and tracking system makes use of a low-power tracking laser, or lidar, and exquisite control systems to hold the high-power laser beam steadily on the target. Other physics considerations include laser-beam/target interaction damage mechanisms and defocusing perturbations (thermal blooming effects), which arise from laser-beam heating of the air column along the beam propagation path.

On another front, significant progress is being made toward development of radiation sources in the blue-green spectral region for use in above-surface to undersea communications systems, an approach made feasible because the optical transmission of seawater is greatest in this spectral region. Alternatively, with an intense blue-green coherent radiation source it might be possible to see into the ocean (like underwater radar) and detect submarines and other submerged objects. Of course, in all the above applications there are fundamental depth limitations imposed by the exponential nature of the optical absorption mechanism.

Additional national-security applications of lasers include manufacturing processes such as laser annealing in microelectronic chip processing, laser isotope separation, remote sensing for meteorological purposes and for detection of toxic agents, sensing of jet-engine combustion and flow parameters, and ring-laser gyroscopes for precision navigation (Figure 12.1).

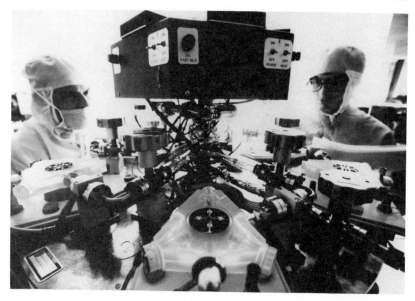

FIGURE 12.1   Ring-laser gyroscope (triangular object in center foreground), shown in volume production facility, serves as an angular rate sensor at the heart of sophisticated, high-accuracy, high-reliability navigation systems. After initial introduction in commercial aircraft, ring-laser gyroscopes are finding widespread application in military systems. (Photograph courtesy of Honeywell, Inc.)

Cyclotron Resonance Masers and Free-Electron Lasers

Cyclotron resonance masers (CRMs), free-electron lasers (FELs), and related devices (such as the gyrotron shown in Figure 12.2) convert the kinetic energy of energetic electron beams directly into coherent electromagnetic radiation. With these schemes, it is possible in principle to achieve high-efficiency, high-power generation of any desired wavelength from microwave (radar) frequencies through millimeter, infrared, and visible wavelengths to the ultraviolet. The national-security implications are thus significant. The CRM is the most highly developed and appears most promising for use in the centimeter to millimeter wavelength regions, where it offers the potential for significant impact on radar technology, especially in countering low radar-profile ''stealth'' technology. The CRM principle was first experimentally demonstrated in the United States, but leadership in its development now rests with the Soviet Union.

Although the FEL offers prospects for coverage of an extremely wide spectral region, it is too early to predict what effects the

FIGURE 12.2 95-GHz gyrotron traveling-wave tube capable of 20-kW peak power at a gain of 30 dB. In operation, the small-diameter portion of the tube is placed in the bore of a superconducting magnet. (Photograph courtesy of Varian Associates.)

technology that it offers will have on national security. But speculations focus on optical countermeasures and perhaps moderate-power directed-energy applications.

Optical Fibers and Integrated Optics

The explosive growth in optical-fiber technology and integrated optics during the past decade is having a profound effect on many systems critical to national security and to the civil sector. A multiplexed-signal optical fiber designed to interconnect a radar transceiver with an aircraft carrier's combat information center some 750 feet away exemplifies the benefits of optical-fiber communication links. The conventional electrical link consists of a bundle of 47 electrical cables carrying 375 separate signal lines and 5 heavy copper ground wires, weighs 7 tons, and costs $1 million to install. In contrast, an optical-fiber link would consist of 7 strands and a spare, would weigh only 15 pounds, and would cost only $30,000 to install. With such marked economy it is reasonable to provide redundant links that follow different routings to ensure integrity even if one path sustains damage.

Accurate Clocks and Relativity Applications

There is a direct relationship in navigation between accuracy in time and accuracy in position. Hence accurate clocks are vital to national-security interests. Physics-devised accurate clocks include the hydrogen maser, the cesium atomic-beam clock, and the optically pumped rubidium cell. Clock accuracy has progressed to such levels that it is now necessary to take relativistic effects into account in advanced navigation systems. To an observer with a clock at low altitude (high gravitational field) a second clock at a higher altitude (lower gravitational field) appears to run faster. A second relativistic effect involves the so-called twin paradox of Einstein's theory of relativity. To a clock in a fixed position, a traveling clock appears to run more slowly. Physicists have recently applied these concepts to timekeeping and time transfer in the so-called Global Positioning System (GPS). This space- and ground-based navigation system, which is now operational, utilizes rubidium clocks in satellites (as illustrated in Figure 12.3) and can pinpoint a user's location to within less than 100 m. If no account were taken of relativistic effects, an error of 12 km would accumulate in the user's location determination in a single day.

Significant progress has been made recently toward new generations of still more accurate clocks based on ingenious laser techniques that cool stored ions to a small fraction of a degree of absolute zero. A new clock based on laser-cooled mercury ions may offer another hundred-fold increase in accuracy. Progress in efforts to cool and store neutral atoms is paving the way for new generations of more accurate clocks for even more accurate navigation and more reliable and efficient communications.

Applications of Ion Implantation

Ion implantation is a technique for introducing dopant atoms into the near-surface region of a material. Application of ion implantation techniques to fabrication of compound semiconductor (for example, GaAs) electronics has been particularly challenging because of the more complex structures of these materials. Impetus has come almost exclusively from the national-security sector, where the higher-speed

FIGURE 12.3 Global Positioning System satellites, such as the one depicted in this artist's rendition, utilize on-board rubidium atomic clocks and theory-of-relativity adjustments to provide routine navigational position fixes accurate to better than 100 m. Omission of relativistic adjustments would lead to an error accumulation of 12 km per day. (Illustration courtesy of Rockwell International.)

performance offered by compound semiconductor electronics is most in demand.

Ion implantation offers benefits in nonelectronic applications as well. Surface-sensitive properties such as fatigue strength and resistance to friction, wear, and corrosion have all been markedly improved. Cutting tools and jet-engine and helicopter rotor bearings have demonstrated significant life enhancements. Although it is expensive, ion implantation can be justified in instances in which enhanced lifetime of a vital component would increase the productive utilization of a high-value system.

Compound-Semiconductor Electronics

Despite extreme technological challenges in comparison with those associated with the simple elemental semiconductors (such as silicon and germanium), progress in the science and application of compound semiconductors (such as GaAs and HgCdTe) has been remarkable during the past decade. Much of this progress has been directly supportive of national-security interests. Compared with the elemental semiconductors, the compound semiconductors, with their more complex chemical nature, are much more difficult to prepare in the extreme purity required for device applications. The process of adding dopants is also made more difficult by the greater chemical complexity of compound semiconductors. Layered configurations of multiple-compound semiconductors for specialized devices require substrate materials and layers with near-perfect match of thermal expansion coefficient and crystal lattice constant.

Overcoming the above difficulties has led to gigabit-rate digital circuits and monolithic microwave oscillators and amplifiers with unexcelled performance for application to electronic countermeasures, communications, and radar systems.

Another unique capability of compound-semiconductor technology is that it permits one to choose (within broad limits) the operating wavelength for a detector or source of electromagnetic radiation. This has led to development of infrared detectors, mosaic focal-plane imaging arrays, light-emitting diodes, and injection lasers tuned specifically to wavelengths of particular interest in applications such as infrared missile seekers, satellite-based imaging infrared surveillance systems, and optical-fiber and integrated optical systems. The ability to fabricate superlattices and quantum-well structures through the use of molecular-beam epitaxy (MBE) techniques makes possible whole new classes of materials and devices. (See, for example, the *Report on Artificially Structured Materials*, National Academy Press, Washington, D.C., 1985.) In one such ultrahigh-speed device, a highly doped low-mobility layer contributes charge carriers to an adjacent high-mobility layer. These fast carriers in the high-mobility layer account for the high operating speed of the device. Compound semiconductors hold great promise for continuing significant contributions in areas of high national-security interest.

Magnetic Bubble Memories

Magnetic bubble memories overcome many of the problems of semiconductor memories and, hence, are finding wide application in both commercial and national-security-related situations. Typical semiconductor random-access memories are volatile. Semiconductor electronically erasable, programmable read-only memories are relatively limited in storage density and, moreover, can be rewritten only a limited number of times.

Magnetic bubble memories are based on small magnetic domains, magnetized in a direction antiparallel to the rest of the medium in which they are immersed. Moreover, it is possible to move or annihilate the bubble domains by application of appropriate guiding magnetic fields. Magnetic-bubble-domain materials can support very high bubble-domain densities (that is, small bubble domains) and can now be prepared with such extreme perfection that defects that might trap or repel bubble domains are absent. Not the least of the challenges has been the development of compatible substrates for support of bubble-domain films. In any event, 64-megabyte magnetic-bubble-domain memories are now available, and it is estimated that in 1986 some $25 million worth of them will be incorporated into military systems.

FUTURE DIRECTIONS

Many of the emerging trends in national-security-related technology pose challenging problems to physics. Three such trends are discussed in this section. The physics subdisciplines that may be expected to affect them should be apparent by inference. However, if history is any guide, revolutionary advances may well emerge from other totally unanticipated directions outside the normal preserve of national-security-related research.

Sensing, Processing, and Deception

In military conflict, adversaries almost invariably attempt to maintain as great a distance as possible from each other, subject to being able to inflict damage with the weapons at their disposal. In any exchange of qualitatively similar weapons the advantage clearly lies with longer-range, better-sighted, and more-intelligent weapons. The challenge of developing long-range capability has largely been met, at least for certain types of weapons, since global ranges have been a reality for more than two decades. On the other hand, the ability of military systems to detect and process data and to make optimum decisions based on this information is often perceived as highly developed; in reality it is in a rudimentary state when compared with the capabilities of its competition, the human brain. Helping to narrow this gap is a task well suited to the methods of physics.

In space, in the atmosphere, and over short distances in seawater, military systems typically see by means of electromagnetic waves (EMWs), whereas over large distances in seawater, military systems see with acoustic waves (AWs). Full exploitation of this capability to see requires that understanding of the generation, propagation, and detection of EMWs and AWs be expanded over ever broader spectral regions. The ability to generate EMWs and AWs efficiently is required for active surveillance approaches such as radar, lidar, and sonar (as well as for communications). Understanding of propagation effects in various media is required if both active and passive surveillance systems are to be utilized optimally. It must be determined what wavelengths of radiation can be most effectively utilized under various environmental conditions and what limitations exist for those conditions, such as ocean waves, currents, and thermoclines; atmospheric rain, clouds, dust, and inversion layers; and ionospheric structure. Better detectors or sensors are required to serve as analogous eyes at optimum wavelengths.

Better seeing alone is not sufficient, however. Many well-sighted military systems provide much more information than their on-board information-processing are capable of interpreting. A television-guided missile is such an example. It carries a television camera that allows it to acquire an optical image of the battle scene. It must relay that image to a remote site, where a human operator receives and interprets the image, deciding which of the various accessible targets should be attacked to inflict maximum loss on the adversary. On-board electronic systems cannot carry out even the image processing. In another example, a radar return signal can tell an operator that an object is located at a particular position, but it cannot identify the object or indicate whether it is friend, neutral, or foe. Yet the radar pulse carries a signature of the object from which it reflected. The situation is in principle similar to the laser-scanned product universal code identification that takes place at the supermarket checkout stand. The way in which these two cases differ is in the time rate and complexity of variation of the return signal, which are much faster for the radar return.

In order for the military identification problem to be solved, extremely fast electronic (or optical) processors of the transmitted information will be required that will be capable of capturing the optical image or radar signal and placing it into memory, decomposing it, and searching for characteristic signatures that are capable of providing an estimate of the object's identity and, if the object is determined to be hostile, activating measures to counter it. Finally, the ideal system should assess the success of the countermeasure, and its cost should be reasonable. The ultimate realization of such affordable speeded-up humanlike capability will rest heavily on progress in computer science and both electronic and optical technologies, areas in which physics often sets the pace of development.

With the establishment of such highly developed processing capabilities, electronic and optical countermeasure and countercountermeasure contests will become not only more complex but perhaps more frequently employed. The energetics involved would appear to dictate this approach. Simply stated, the energetically most efficient means of countering an "intelligent" weapon is to transmit to its sensors and central nervous systems a relatively low-power signal appropriately coded to confuse it to such a degree that it cannot accomplish its mission. Clearly, in such contests the advantage will be with those who possess the greater mastery of signal processing and interpretation.

Directed-Energy Weapons

In general, the most efficient weapon places the right amount of energy of the most effective form in the most effective place at the most effective time at the least cost. Viewed in these terms, traditional methods of energy delivery by guns and missiles, though remarkably effective, are slow and relatively inefficient. This accounts for the interest in directed-energy weapons such as high-power lasers, high-power microwave sources, and energetic-particle-beam accelerators. These approaches offer the possibility of transmitting significant amounts of energy from source to target at or near the speed of light and with relatively high efficiency. In space, EMWs and energetic neutral particles (atoms) propagate in straight lines with only normal spreading losses. (Charged particles will be deflected by the Earth's magnetic field and will tend to disperse because of mutual repulsion.) In the atmosphere EMWs experience varying amounts of absorption depending on wavelength and are deflected by atmospheric inhomogeneities, but these perturbations do not present insuperable barriers. The use of energetic particle beams in the atmosphere is much more speculative, inasmuch as absorption is high and the perturbing effects of the Earth's magnetic field are unavoidable.

All factors considered, it is likely that directed-energy weapons ultimately can be developed that will be capable of fulfilling some critical military missions. Although their effect could be revolutionary, the time scale is highly conjectural. Such systems may be land, sea, air, or space based. Some may even use Earth-based lasers and space-based mirrors. The earliest operational systems will doubtless be designed for short-range, modest-power engagements aimed at whatever sensor or electronic, optical, or control system might be most vulnerable. Ultimately the vital components of intended targets will be so hardened against such attack that very-high-power directed-energy systems capable of inflicting structural damage from great distances will be required. In all phases of these developments, from laser and accelerator sources to beam interaction with the propagation medium to beam-target interaction to target-hardening countermeasures to damage assessment, physics will have to play a vital and central role.

Low-Observables Technology

The term low observables is used to designate approaches designed to render military systems less easily detectable. The goal may be to maintain sustained concealment, to reduce the range at which detection is possible, or to reduce the radar cross section of a high-value unit and render it indistinguishable from low-value units in the same general area. The term low observables may refer to a great variety of approaches: traditional camouflage, acoustic silencing of noisy components on board submarines, or even vehicular designs that minimize radar reflection glint. Finally, it may include active cancellation measures in which a vehicle first senses both its own emanations and surveillance signals directed against it and then generates signals of such phase, frequency, and amplitude as to compensate for the aforementioned signals and thus appear indistinguishable from the background. Although it may, in principle, be possible to approximate this ideal for a particular wavelength region, it is clearly not practical across a broad-wavelength spectrum.

Countermeasures against low-observables technology include more-sensitive detectors, more-sophisticated signal-processing approaches, increased radiated power levels of active surveillance systems, and frequency diverse or multimode surveillance systems such as radar and lidar together. The extent to which the above low-observables measures and countermeasures may prove operationally practical remains to be seen. But, in any event, the scientific and technical issues involved will require the continued involvement of physics research and development.

PHYSICS AND ARMS CONTROL

Strategies for achieving national security are both complex and multifaceted. In general, these strategies place emphasis not only on advances in armament but also on arms-control agreements. Although much of the focus of this chapter has been on advances in armaments, this should not be interpreted as implying that every advance in armament by the United States necessarily results in increased national security. Indeed, an asymmetric advance in armament may be either stabilizing or destabilizing, depending on a variety of factors. For example, one nation may seek to exploit an advance in armament either as a means to enforce stability or as a means to pursue aggression, while the nation not possessing the advance may be so intimidated as to accept stability or, alternatively, be compelled to undertake immediate aggressive action to pre-empt growth of asymmetry. Moreover, advances that appear to be symmetric militarily may be asymmetric economically and hence destabilizing.

In arms-control agreements, potential adversaries negotiate limitations on the development, production, and deployment of arms in attempts to reduce tension and to achieve both greater security and lower burden of armament. The negotiation, implementation, and execution of arms-control agreements are highly complex and involve both political and technical considerations. During the Eisenhower administration, physicists serving on the President's Science Advisory Committee made significant contributions to the institutionalization of the arms control process within the federal government. Physicists have since provided vital technical expertise to assist negotiators in framing agreements that are susceptible to verification and that, in the final analysis, are judged to be in the best interests of the United States. After consummation of agreements, physicists have fulfilled essential roles in the development, establishment, and operation of compliance verification systems. Particularly noteworthy were contributions by physicists to the technology of monitoring nuclear test explosions in connection with the Limited Test Ban Treaty of 1963.

Satellite technology, which is reliant on sophisticated developments in physics for instrumentation for receiving and analyzing signals over a wide range of the electromagnetic spectrum, is an essential ingredient of systems used to monitor compliance to arms-control agreements. The integrity of such satellites in peacetime has been guaranteed by the Strategic Arms Limitation Treaty. Current discussions of space-based nuclear defense systems and of potential agreements for their control inevitably involve a plethora of fundamental physical problems in a new context. The sheer magnitude of the implied commitments of national resources and their effect on the security of mankind demand the most thorough consideration of the scientific fundamentals at every stage.

ENHANCING THE NATIONAL-SECURITY/PHYSICS INTERFACE

Remarkable and revolutionary impacts of physics on national security have been chronicled above. None of these developments could have been anticipated at the dawn of the twentieth century. Moreover, in many instances the seminal physics research efforts that made them possible (in stark contrast to the development efforts that made them operational) were undertaken in peacetime with no realization that a result of great potential impact on national security might emerge. There is a cautionary lesson therein. Invention and development based on physics can flourish only insofar as the knowledge bank in physics is rich, diverse, and expanding. This perspective appears to be understood at the highest level of the defense establishment. In the February 1983 issue of *Defense 83*, the Secretary of Defense wrote that

. . . we must also recognize that our current advantages come principally from the momentum in basic technologies we developed during the 1950's and 1960's, the "golden period" of United States scientific exploration. We face the danger of losing our edge because we have not adequately replenished the reservoir of scientific concepts and knowledge to nourish future technologies during subsequent years of fiscal neglect of defense research and development. Given these circumstances, we must systematically replenish that scientific reservoir, using the unique and diverse strengths of the United States scientific community, to insure that the technologies necessary to sustain our national defense in the decades ahead will be second to none.

While the defense agencies and their contractors are major users of knowledge in physics, they represent only one of the many sectors that contribute to and utilize that store of knowledge. Each sector profits from the contributions of the others to that store of knowledge, and each sector profits from the sustained vitality of the entire physics community. If, for example, the National Science Foundation were to abandon physics research, all users, including the national-security sector, would either have to fill the void or suffer serious consequences.

Advances in physics are almost never so narrow as to be of value to only a single using sector. To attempt to categorize subfields of physics strictly as being exclusively Department of Defense (DOD) type, or Department of Energy type, or National Aeronautics and Space Administration type, for example, is to assume great risk. Indeed, if agencies restrict all their efforts to activities that can be categorized as theirs exclusively, then their efforts will most likely be exclusively developmental, and their couplings with research foundations will most likely be compromised. Likewise, to assign long-range research exclusively to the National Science Foundation while assigning applied or short-range research exclusively to the so-called mission-oriented agencies would do great damage to the essential information transfer across the interface between long-range and short-range research. Succinctly stated, both long-range and short-range physics research should be vital components of the programs of advanced-technology agencies and corporations. Although we concentrate here on the physics/national-security interface, these issues are common to all long-range scientific research.

To assess the interface between national security and physics, it is thus important to distinguish between long-range and short-range physics research. Many factors suggest that the interface between short-range physics and national security is in relatively good condition. Physicists are widely and appropriately employed in vital short-range defense programs. Examples cited above show their profound impact. This part of the system, in which physicists draw on the knowledge base for application to near-term development projects, appears to be working well. There should, however, be concern for the currency and ultimate effectiveness of such applied physicists unless they are well connected with the more basically oriented physics community and unless the defense-contract research agencies are similarly optimally coupled with the basic physics community. Two types of evidence suggest that such a troubling circumstance has indeed developed: funding patterns and publishing patterns for defense-agency-supported research.

Corrected for inflation, the 1983 combined U.S. Army, Navy, and Air Force funding level for physics research was 30 percent below the 1969 level. Moreover, the *integrated effort* over that 15-year period fell 4.7 years behind that which would have been the case had the actual effort been maintained constant at the 1969 level. Based on defense-physics research funding as a percentage of gross national product, the lag amounts to 6 years. By these criteria, it is apparent that the interface between physics and national security has been severely weakened. Moreover, this weakening of the interface has been compounded by two other factors: a shift by the defense-research agencies away from long-range research and toward shorter-range research or even development and the concomitant widespread abandonment of long-range physics research programs by U.S. industrial firms, both defense-oriented and commercial. Evidence bearing on these factors is examined below as a prologue for discussion of possible approaches for revitalizing the interface between national security and long-range physics research.

The most prominent early sign of erosion of the long-range research outlook was the enactment of the Mansfield Amendment (1970), which was seen as a detriment to longer-range programs. Defense-agency research funding was decreased sharply in real terms, and many longer-range university and defense laboratory physics efforts were discontinued. The uncertainties of defense-agency research policies and funding led university researchers to establish long-term contract and grant relationships with nondefense agencies whose policies were judged to be more congenial. At the same time, the defense industry abruptly disengaged from long-range research. Scientists were suddenly unemployed, and university science enrollments plummeted. Now, some 15 years later, several features have dissipated: the 2-year life of the Mansfield Amendment is far in the past; defense, aerospace, and civil sector industries are thriving; and physicists are again in demand, at least for short-range program activity. But the weakened connections between national security on the one hand and academe and long-range physics research on the other have not been revitalized.

Dramatic changes in publication practices over the same period offer additional evidence of these weakened conditions. The frequency of publication of defense-agency-funded research results in *Physical Review Letters* (PRL) is shown in Figure 12.4.

The sharp decrease beginning after 1968 coincides with the enactment of the Mansfield Amendment, and, although that amendment was in effect only until 1971, the depressed publication rate has persisted. PRL provides rapid publication of brief reports on work at the forefront of basic physics, covering the entire breadth of physics. Papers published in this journal are generally believed to be of exceptionally high quality, are representative of the most rapidly developing aspects of basic physics, and must be so significant as to be deserving of rapid publication. Physicists from around the world compete vigorously to have their work appear first in PRL. Approximately 39 percent of those who succeeded in 1983 were U.S. university faculty members, down from a peak of 55 percent in 1967. (As an aside, foreign-based physicists accounted for approximately 36 percent in 1983, up from 17 percent in 1967.)

The data shown in Figure 12.4 imply a decoupling of the national-security establishment from the intellectual community, largely university based, which publishes in PRL and from which the most revolutionary advances in physics typically emerge. Also troubling are the PRL publications data for the in-house defense laboratories. During the

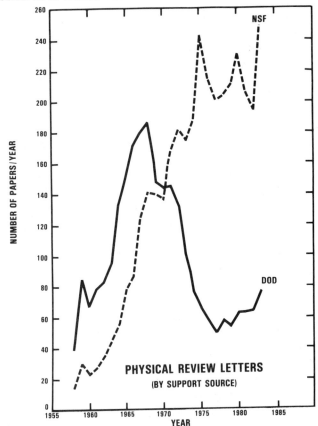

FIGURE 12.4   Annual number of PRL articles describing DOD-funded research (solid line) and NSF-funded research (dashed line) for the years 1958 through 1983.

period 1958 through 1983 the U.S. Navy, Army, and Air Force laboratories accounted, respectively, for only 1.2 percent, 0.1 percent, and 0.1 percent of all PRL papers. Among the 70 or so defense-oriented laboratories, only the Naval Research Laboratory appears to maintain significant activity at the forefront of physics. Equally disturbing is the fact that, with the exception of AT&T Bell Laboratories and IBM, U.S. industrial laboratories account for only a few tens of papers per year in PRL. This, despite the fact that the so-called independent research and development (IR&D) costs recoverable by industrial defense contractors from DOD amounted to $1.4 billion in fiscal year 1983. Thus, although defense (and civil sector) industrial organizations spend heavily on research and development, their ability to capitalize rapidly and efficiently on significant advances that occur in basic physics is questionable.

These circumstances are now deeply ingrained in the very fiber of the defense and industrial research and development establishments and cannot be attributed simply to the Mansfield Amendment, which appears to have been only a single facet of a much larger movement in government and industry away from the risk of long-range research. This movement appears, in turn, to have been symptomatic of a still more general tendency in broad realms of endeavor to focus on short-range problems and to sacrifice long-range capability.

There are encouraging signs that government and business leaders alike have recognized this general problem and are working diligently to overcome it. The substantial recent increase in National Science Foundation funding approved by Congress is one such encouraging sign. The negotiation of some rather substantial and long-time-period contractual research arrangements between selected universities and some forward-looking corporations is another. *But the problem identified above, having to do with the interface between the defense establishment and long-range physics research, has not been adequately addressed.* Constructive measures that should be instituted to deal with it are discussed below.

Physicists themselves should play an active role in the solution of this problem. However, because there now exists a new generation of physicists who have had no historical ties with national security, the re-establishment of a strong interface will not be easily accomplished. Nevertheless, physicists should be diligent, articulate, even persistent in marketing long-range physics research proposals directed at all sectors of the defense establishment, namely, the defense research agencies, defense laboratories, and industrial defense contractors. Physicists should also seek to serve on government committees and panels that influence defense policies and programs, and they should seek roles as consultants or board or panel members for defense corporations. In short, physicists themselves should take initiatives to strengthen the interface between defense and long-range physics and thus contribute their special expertise to the formulation of national-security policies and measures.

Defense contractors should rebuild their in-house long-range research programs by addition of long-range-oriented staff, reassignment of existing staff to longer-range projects, and sabbaticals for existing staff to perform research on university campuses. This will foster more-effective communication with, and utilization of, university basic physics resources. In the long term, the greatest value to the corporation will lie more in the cumulative effect of its association with, or access to, universities as a whole than in its association at any given time with particular faculty members. Because high-level defense officials have encouraged industry-university associations such as those described above, there is assurance that their cost will be allowable under IR&D provisions.

Legislative and executive branches of the government should take actions to expand the fraction of defense research and development funding that is devoted to long-range research. *Safeguards should be established to ensure that activities supported with funds appropriate for defense physics research are not diverted from long-range research purposes.* The DOD technological program needs the close contact with research in many areas of basic science including physics that can be achieved best by participation in a share of funding of long-range basic research commensurate with its overwhelming requirements for research and development in the nation.

The balance between short-range and long-range efforts within the $1.4 billion per year defense-contractor IR&D program should be of equal concern to government and defense industry officials. Under the IR&D program, corporations that are awarded large defense contracts are reimbursed for so-called independent research and development tasks, which, although not explicit parts of particular contracts, are nonetheless essential if these corporations are to maintain their strong technology-based positions. For the past 15 years, however, there has been little evidence of long-range physics research activity in defense contractor IR&D programs, a circumstance that is neither in the best interests of national security nor of the defense contractors themselves. Defense contractors are strongly urged to re-establish substantive ties with the long-range physics research community through measures such as those discussed above. Legislators and government executives should establish policies and measures that are supportive of this endeavor.

For university research programs, there exists no discretionary funding analogous to the IR&D program and to the in-house defense laboratories through the independent research and independent exploratory development programs. Yet the universities perform about 50 percent of DOD basic research, provide expert professional advice, and serve as the source of trained scientific and technical personnel for the entire national-security establishment. It would appear to be equally in the interest of national security that universities also be allocated discretionary funding to perform research beyond the constraints typically imposed by explicit, contractual arrangements. Accordingly, it is proposed here that Congress appropriate additional funds specifically for *university* IR&D programs to be administered much as industrial IR&D programs are administered along directions pertinent to national security. Such additional funding could be particularly well directed to, among other problems, the widely recognized one of providing seed money to initiate the research careers of young, new faculty members. It is important that the peer-review process be incorporated into any such initiative in order to ensure the highest level of scientific quality.

Taken altogether, the recommendations set forth above provide an agenda for progress in the utilization of physics in particular, and science in general, in the interests of national security. Progress toward realization of these goals will require the dedicated and cooperative efforts of all concerned—the public, government, industry (both defense and nondefense), academe, and, above all, physicists themselves.

13

Medical Applications of Physics

INTRODUCTION

A strong relationship has existed between physics and medicine, in part because of the classical tradition of the physician-scientist. In that tradition the possibility of medical application of a new scientific development is quickly appraised. For example, the first formal report by Röntgen of his discovery of x rays was made to a medical society. Despite the sophistication and specialization demanded by modern physics and modern medicine, the fields maintain a strong interaction. Indeed, modern medicine owes much to modern physics.

The contributions of physics to medicine can be described at various levels. Much could be written about the impacts of basic discoveries of physics on medicine, which is, among other descriptors, an applied science. One example is the discovery of x-ray diffraction, which led to the knowledge of the three-dimensional structures of biomolecules that was crucial to understanding biology at the molecular level with important impact on such areas as pharmacology and genetic diseases. Much could also be written to trace discoveries in physics to implementations in devices that facilitate medical practice. What is by now the classic example is that of the transistor. Medical practice utilizes a broad range of devices that contain microelectronics. In fact, none of the modern diagnostic imaging modalities, such as x-ray computed tomography, could be implemented without microelectronics technology, nor could the implanted cardiac pacemaker.

Neither of these kinds of contribution of physics to medicine will be addressed in any detail here. Such contributions are mentioned in other chapters of this volume. What is highlighted here are some of the developments in medicine in which physics and physicists played a rather direct role. Much of the discussion concerns noninvasive diagnostic devices that provide internal body images. These imaging modalities, such as x-ray computed tomography, magnetic-resonance imaging, and ultrasonic imaging, have been revolutionizing medical practice. Another major area described is the utilization of optical devices in medical diagnosis, patient monitoring, and therapy. The use of lasers in surgical procedures is expanding rapidly from the initial uses in ophthalmology. As the ability for remote control of fiber-optics delivery systems improves, minimally invasive laser procedures will replace some procedures now described as major surgery.

Much of the discussion centers on developments that have occurred in the past two decades. Many of these developments, especially those concerned with radioisotopes, derive from connections between physics and medicine originally made toward the end of World War II. Some of this history is traced. In addition, some devices now in early development stages are described to provide a picture of continuing trends and a glimpse of the future. It appears that trends extrapolate to a future that may not be badly represented by the medical practice found in science fiction where a hand-held device waved over the patient gives within seconds a detailed diagnosis on a visual display. It is much more certain to say that trends indicate that medical practice will be significantly changed a decade from now because of both indirect and direct applications of discoveries and inventions in physics.

RADIOLOGY

If one were to look within a typical community hospital for personnel called physicists, one would find them in the radiology department. It has become practice for the work of calibrating radiological instruments, mapping therapeutic radiation doses, and protecting patients and staff from radiation hazard to be carried out by graduate physicists specializing in what is called medical physics. This practice resulted from the direct interaction between physicists and physicians stemming from early work on x rays and burgeoning with the advances in radiation therapy and diagnostic nuclear medicine that emerged from the nuclear-physics efforts of the 1940s. The need for more-accurate knowledge of the attenuation of various radiations passing through patients under radiation therapy led medical physicists to the development, in the 1970s, of x-ray computed tomography, with its ability to image soft tissues with unprecedented contrast. The success of x-ray computed tomography stimulated the development of a variety of advanced imaging modalities (such as positron emission tomography and magnetic resonance imaging). The revolutionary nature of these developments is indicated by the fact that departments of diagnostic radiology are changing their titles (to departments of diagnostic imaging, for example).

Diagnostic Radiography

X-ray imaging was at first confined to the diagnosis and treatment of bone fractures and to the localization of foreign bodies. Its chief limitation was its failure to produce a contrast between the images of adjacent body structures because of the similarity between the densities of adjacent soft tissues. Conventional x-ray radiography requires a density differential of a few percent to achieve conveniently detected contrast. However, even early instrumentation found extended utility through the use of contrast agents, either x-ray-opaque substances with elements of high atomic number (such as barium and iodine) or gases (such as air) with much lower density than tissue. Immense benefit has come from the use of barium sulfate in the examination of the gastrointestinal tract. This technique continues to be refined in significant ways concomitant with improvements in real-time detection and display technologies. Iodine-containing compounds continue to be developed for examination of the urinary tract, the biliary tract, the heart, and various blood vessels. This approach finds extreme sophistication in assessments of the degree of occlusion of renal, carotid, and coronary arteries. These methodologies derived from advances in condensed-matter physics that led, starting in the 1950s, to electronic detection of x rays, image intensifiers, improved cathode-ray-tube (CRT) displays, and electronic recording media. Electronic recording allows for convenient digitization of the image and computer-aided image processing. In digital subtraction angiography, recently introduced into use, images are viewed before and during introduction of contrast agent, and a difference image is obtained. This removes shadows not caused by the contrast agent, allows for better image definition, and requires the use of much less contrast agent. In an early experimental phase is another scheme for enhancing the image of an iodine-containing contrast agent. Monoenergetic x-ray beams from a synchrotron source with energies just above and just below the iodine *K* edge are impinged simultaneously on the subject. Subtraction of the two transmitted intensities removes all shadow-forming structures except for the contrast agent.

Various schemes for the totally electronic radiology department are under development. Both direct digital recording area detectors and indirect systems have been introduced. In one indirect system the x-ray image is stored on a flexible plate of europium-activated barium fluorobromide, which takes the place of the usual photographic film. The reusable plate is scanned with a beam from a helium-neon laser to induce electron-hole-recombination luminescence, which is digitized and stored electronically.

Coupled with careful attention to x-ray energies and exposures, conventional film-based radiography continues to provide a moving target for electronic recording in such areas as mammography. However, the expected savings in photographic film and, more important, in labor for storage and retrieval of films, in physician time and hospital-stay time, and in the increased efficiency that may result from image processing is driving the field toward electronic recording. The film-less, totally digital radiology department is not far in the future.

Isotopes and Nuclear Medicine

The invention of the cyclotron in the early 1930s brought the ability to make radioactive isotopes in quantities that stimulated uses in biological and medical research. The development of nuclear reactors in the 1950s led to neutron-produced radioisotopes in quantity and to their routine application in medicine. It is not an overstatement to say that little of the knowledge of molecular biology, metabolism, and physiology gained over the past four decades would have been obtained without isotope tracer techniques. The discussion here is confined to the use of radionuclides for in vivo medical diagnostics except to mention radioimmunoassay (RIA). RIA employs the coupling of antigen with specific antibody. The antigen or antibody is labeled with a radioactive nuclide, usually iodine-123, permitting measurement of the concentration of virtually any substance of biologic interest, often with unparalleled sensitivity. RIA methods are adaptable to automation and are the methods of choice for in vitro diagnostic assays of peptide and steroid hormones, for example.

The use of radioactive substances in medical diagnosis is referred to today as nuclear medicine. In a typical examination a radioactive tracer is introduced, usually intravenously, and the distribution of the radio-nuclide within the body or a targeted organ is displayed in a series of two-dimensional images. The imaging is based on the emission of gamma rays by the radionuclide that pass out of the body and are recorded by a scintillation camera.

In the pioneering experiments in 1939, Hamilton and Soley observed the rapidity with which injected radioiodide is accumulated in the thyroid glands of rabbits and confirmed the observation in patients with hyperthyroidism by placing a Geiger counter against the neck. The radioiodine accumulation test soon became a routine assay of thyroid status. Since then radionuclide tracing and imaging have been applied to diagnostic studies of nearly every organ of the body. Major advances in the past two decades have been fueled by vast improvements in scintillation-camera (gamma-camera) technology and data-processing capabilities and by the continuous development of highly specific tracers. The advances in instrumentation depended on both direct and indirect input from physics.

It is worthwhile to mention a few of the routine assays that use nuclear medicine. Currently ^{123}I sodium iodide is used for thyroid function imaging. Several different ^{99m}Tc-containing phosphates and phosphonates are used for bone scans, primarily for detection of metastatic disease in bone or primary bone tumor. A variety of pulmonary disorders can be assessed by radionuclide imaging. For example, ventilation can be assessed after single inhalation of ^{133}Xe or one of a variety of radioactive aerosols. Cardiac assessments include evaluation of myocardial infarct through imaging of ^{99m}Tc pyrophosphate, for example, which localizes in damaged myocardial cells. The complete list of currently used diagnostic radiopharmaceuticals and associated procedures is long and continues to grow. With respect to tracers, the specificity that can be built in by the synthetic chemist may soon be surpassed through the use of radioactively labeled specific (monoclonal) antibodies. With continued improvements in detector cameras and further development of tomographic approaches (see below), nuclear medicine, a rather direct result of discoveries in physics, will continue to be an important diagnostic methodology.

X-Ray Computed Tomography

Computerized x-ray transmission tomography, commonly called CT, produces images of anatomical planes (slices) through the body with density resolution of about 0.5 percent such that soft tissues are imaged. The development of CT during the 1970s is the most significant advance in medical imaging since the discovery of x rays and has pushed medical imaging further into the realm of advanced physics and engineering. CT served as the prototype application of the general reconstruction problem (see below) for medical imaging.

Out of a need to obtain x-ray attenuation coefficients to correct for body inhomogeneities in planning radiotherapy regimens, A. M. Cormack, a physicist, formulated the basis for CT in 1956. He pointed out that a cross-sectional matrix of coefficients could be determined if measurements of x-ray transmission were obtained at many angles and projections through the body and that a gray-scale image of internal anatomy could be reconstructed from such coefficients. He performed a number of successful demonstration experiments and published reports in the physics literature.

Following Cormack's work, Hounsfield constructed the first CT prototypes for clinical applications. The first CT scanner for the head used one x-ray beam and one NaI detector and required 5 min to acquire data followed by 20 min for image reconstruction. Present instruments employ multiple beams and detectors and record data for an image in 2 s. Huge array processors perform image reconstruction in a few seconds. Images of axial planes are obtained in series. Images of coronal and sagittal planes can be constructed from the axial images. CT provides a gray-scale range of 2000 (11 bits), which is beyond the dynamic range of electronic displays and visual perception. Because the computer stores all the data it is possible to recall the image in various perceptible gray-scale windows (levels of contrast), a great advantage for interpretive reading.

CT scanning was first applied to the head and proved quickly to be of such value in demonstrating cerebral abnormalities that it has virtually replaced invasive procedures. CT has been found to be useful in examination of the thorax and pleura and particularly useful for examining the abdomen. CT scanning has also found great value in radiotherapy treatment planning, the original goal of Cormack. Improvements in sources and detectors and in data processing continue to reduce scan time and dose and to increase diagnostic utility. The future should bring scanners in which electron beams streak across crescent-shaped sources to produce focused x rays, eliminating moving parts and permitting ultrafast (<0.1-s) scans. Portable CT scanners that use isotopes such as gadolinium-153 as sources are under development.

The success of x-ray CT is largely responsible for the renewed interest in the tomographic reconstruction problem. The problem is to determine the internal structure of an object without cutting into the object. Interactions between the internal structure and some probe lead

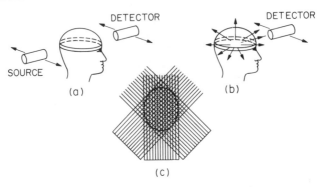

FIGURE 13.1   Tomographic imaging: (a) transmission mode, and (b) emission mode. (c) A typical scanning pattern of linear and angular increments.

to a measured profile. An image of the internal structure must then be reconstructed from this profile. This general approach was well described as early as 1917 by Radon, and is depicted in Figure 13.1. Physicists and mathematicians have worked on the reconstruction problem in fields from astrophysics to materials sciences and have used various probes, including x rays, visible light, microwaves, electrons, sound waves, and magnetic resonance signals. X-ray CT is an example of reconstruction tomography in the transmission mode. Ultrasound CT is similar in approach. In emission tomography, exemplified by positron emission tomography and gamma-emission tomography, one images the internal distribution of a radiation-emitting probe. Magnetic resonance imaging (see below) represents still another variety of reconstruction imaging.

Positron Emission Tomography

In positron emission tomography (PET) the distribution of a positron source within the body is reconstructed from the profile of gamma rays, the products of positron annihilation, detected outside the body, as shown in Figure 13.1(b). Sufficient quantities of positron-emitting isotopes of carbon, oxygen, nitrogen, and fluorine can be produced with the use of a small cyclotron to permit manufacture of adequate amounts of labeled metabolites or analogs for use in diagnostic PET. Because they are short lived (half-lives <1 h), the isotopes must be produced and converted to diagnostic probes on site. It appears that this may not be too costly or unwieldy for most major medical centers to undertake.

PET allows one to image sites of specific biochemical activity. Thus, abnormalities in metabolism, in addition to alterations in anatomical features, can be assessed. This approach holds the promise not only for new and improved diagnosis and for ways to follow therapy but also for increased understanding of factors that influence metabolism.

While PET has been used to examine various organs, the most impressive results to date have to do with brain metabolism probed with labeled 2-deoxyglucose. Like glucose, this analog is taken up by neurons at a rate that increases with the metabolic activity of the cells. Unlike glucose, 2-deoxyglucose is not fully metabolized and accumulates in the cells. Thus, the distribution of 2-deoxyglucose maps the relative metabolic activities of brain regions. Brain tumors can be detected in this way. More exciting, PET is revealing much about locations of brain functions and deviations in distribution patterns with various central-nervous-system disorders. In studies of brain function, PET reveals, for example, that a different 2-deoxyglucose distribution is evoked by stimulation of the visual system from that evoked by stimulation of the auditory system.

PET is limited by the scope of probe structures that can be synthesized rapidly from the labeled starting materials. However, one of its advantages is high sensitivity, and only trace amounts of probes are required.

ULTRASONICS

Unlike other noninvasive methods for imaging internal structure, ultrasonic imaging and allied measurements are commonly used directly by physicians other than radiologists, for example, cardiologists and obstetricians. This is so because no ionizing radiation is involved (this is not to say that ultrasound is perfectly harmless), the instrumentation is relatively inexpensive, and the approach has yielded a variety of specific applications that require different instrumental features.

The concepts and practice of medical ultrasonics derive directly from sonar (sound navigation and ranging), which originates in the work of the French physicist Paul Langevin on piezoelectric transducers in 1917. Spurred by military applications as well as applications in devices for communications, work on piezoelectrics increased dramatically during and after World War II. Advances in fundamental understanding of the physics and chemistry of solid-state materials during the 1950s led to the discovery of superior piezoelectrics for transducers, such as the now commonly used lead zirconate titanate. There was a steady increase in research on the attenuation of sonic energy in condensed materials and on applications of ultrasound in medicine during the 1960s. But it was the availability of transducer arrays and of powerful and inexpensive control, computational, and display electronics in the 1970s that allowed a profusion of diagnostic ultrasound instruments to be devised.

In medical ultrasonics, pulses of ultrasonic (1-20 MHz) waves are launched from a piezoelectric transducer(s) into the body from the skin surface, and reflected wave pulses are detected by the same transducer(s). The amplitudes, time delays, and frequencies of the reflected waves (echoes) are analyzed for information concerning the locations, densities, and motional velocities of parts within the body. Reflections occur at sound-impedance discontinuities. A variety of scan protocols and signal displays are in use. The simple pulse-echo single-transducer A scope displays echo amplitude versus echo time, which corresponds to positions of echo-producing targets along the ultrasonic beam within the patient. The same information may be displayed as CRT brightness versus time (brightness increases with echo amplitude)—a B scan. (The A and B terminology is analogous to that of display modes used in radar.) Several types of two-dimensional B scanner are in routine clinical use: the static B scanner, the water-bath coupled scanner, and the real-time scanner. In these, information about the position of the ultrasonic beam path is acquired automatically, and scanning (manual or automatic) leads to tissue maps (images) in the scan plane. In the most sophisticated instruments scanning is done by electronic steering of the ultrasonic beams from transducer arrays.

The Doppler shift of ultrasonic frequency is widely used for extremely valuable assessments of moving structures (blood, heart walls, heart valves). The Doppler shift of a continuous-wave ultrasound beam is at present the most common way to assess the velocity of a moving body element or to locate flowing blood. This is being replaced by so-called pulsed Doppler systems, which combine the ranging and imaging capabilities of pulse-echo methods with the velocity measurements of Doppler.

Ultrasound is extremely valuable in angiology. Many of the abdominal blood vessels, including the aorta, the inferior vena cava, and portal and hepatic veins can be visualized and aneurysms and thrombi well demonstrated. High-resolution images of leg and neck (carotid) arteries can be obtained. These images in combination with blood-velocity measurements are used to assess occlusion in these vessels. In cardiology the main application of ultrasound is for cardiac valve studies, the mitral valve being the easiest to evaluate because of its position. A variety of cardiac abnormalities, including congenital heart disease, hypertrophic cardiomyopathy, and septal defects, can be assessed. The advent of high-resolution, real-time, two-dimensional scanners has made ultrasound of particular value in gastroenterology. Liquid-filled and solid lesions of the liver, such as metastases, can usually be identified, as can cirrhosis and fatty liver. Gallstones are easily detected, and the kidneys, pancreas, and spleen can be well imaged. Uses of ultrasound in obstetrics include assessment of fetal maturity, detection of fetal abnormalities such as spina bifida, and guiding of amniotic-fluid aspirations.

There is every indication that substantial advances in medical ultrasonics will continue. Although advances in transducer technology provide some of the drive, it is the combination of real-time signal-processing technologies with advances in the physics of interactions of ultrasonic waves with body components that is the chief impetus. Pulsed Doppler ultrasound synchronized with electrocardiographic signals is being developed into systems for analysis of atherosclerosis throughout the arterial system. High-resolution real-time scanners will eventually be capable of displaying the image of coronary arteries. Miniaturized transducers are being developed for use by surgeons for guidance in many operations and for use with catheters. With continued advances in display technology it may not be long before hand-held portable real-time scanners are developed.

The first major therapeutic use of ultrasonics, the fragmentation of kidney stones, has appeared recently. In the invasive ultrasonic lithotriptor, high-frequency ultrasound is conveyed to the stone through a hollow metal tube run through the working channel of the nephroscope. The noninvasive lithotriptor aims high-frequency ultrasound beams at renal calculi through a water bath in which the patient is immersed. Early clinical trials indicate that the noninvasive method is efficacious in about 60 percent of cases.

NUCLEAR MAGNETIC RESONANCE

In 1984, the U.S. Food and Drug Administration approved the sale of nuclear magnetic resonance imaging systems for clinical use, marking the beginning of yet another revolution in diagnostic medicine. Appearing just a few years after x-ray CT, magnetic resonance imaging (MRI) can provide impressively clear images of soft tissues and flowing blood and the possibility for chemical analysis of internal structures, as shown in Figure 13.2. MRI is also exceptionally safe; no ionizing radiation is involved, and the amount of energy dissipated within the body appears to be insignificant.

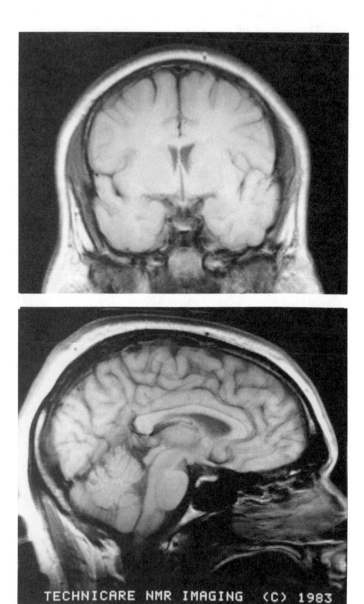

FIGURE 13.2 Proton MRI of coronal (*top*) and sagittal (*bottom*) slices of normal cranial anatomy. The images, from a Technicare 1.5-tesla Teslacon™ magnetic resonance system, were obtained using a spin-echo pulse technique.

Nuclear magnetic resonance (NMR) was conceived by physicists E. M. Purcell and F. Bloch in 1946 as a way to determine magnetic properties of atomic nuclei. Nuclei with magnetic moments, for example, ^1H, ^{23}Na, and ^{31}P, align in an applied magnetic field to produce two energy states. Transitions induced between such states is detected in NMR. Because characteristics of NMR transitions are sensitive to the interactions of the examined nuclei with surrounding atoms, NMR provides a wealth of information about structure and dynamics in condensed matter. NMR quickly became an indispensable tool for structural chemistry. With this as a driving force for advancing instrumentation, analytical NMR has the capacity today of resolving the three-dimensional structures of even large biomolecules. Increases in sensitivity and in specimen size afforded by the high-field magnets made possible because of superconducting alloys, another major discovery of physics of recent times, provided the means for studies of metabolism in living cells and perfused organs beginning about 1970. These studies and work under way on living small animals are already yielding new knowledge and revising long-standing conclusions about organ and muscle metabolism, physiology, and drug action. Physics and physicists played key roles throughout these developments.

R. Damadian suggested in 1972 that NMR could be used as an in vivo tool. He had observed differences between the water proton magnetic-resonance relaxation times of cancerous tissue and normal tissue. He proposed to make measurements at localized points in the body defined by degrading the static magnetic field everywhere except at the point of interest. However, it is the concept of using uniform magnetic-field gradients, first promulgated in 1973 by P. C. Lauterbur in the United States followed shortly by P. Mansfield and P. K. Grannell in England, that makes MRI possible. The resonance frequency of the nuclear-spin transition depends on the magnitude of the external applied field as well as on the local environment of the nuclei of interest. For example, protons in different molecules have different resonance positions for the same external field. Most of the protons in soft tissue are in water molecules, and the single resonance line of water dominates the proton NMR spectrum of the body. If one ignores all but the water protons, position in a spatially distributed specimen can be encoded in the resonance frequency in a simple way for the case of a linear applied-magnetic-field gradient.

MRI instrumentation can be configured to receive data from a single volume element within the sample, from a line of elements, from a plane, or from a large volume all at once. Two-dimensional images in any orientation can be reconstructed from the data. In proton MRI contrast is provided not only by the differences in proton (water) concentrations in different tissues but also by local differences in resonance relaxation times. Flowing blood is revealed in high contrast in MRI, because the motion provides a large effective change in relaxation rate. Bone is essentially transparent in MRI, with the result that excellent images of the spinal cord can be obtained. Currently, proton MRI instruments are capable of spatial resolution of the order of 1 mm in regions of high contrast.

Motion in the thorax and abdomen during the several minutes required for scanning degrades MRI images so, at this time, clinical use of MRI has been more or less restricted to examination of the head, spine, pelvic region, and extremities. In these areas extraordinarily detailed images can be obtained. Synchronization with electrocardiographic signals has permitted excellent imaging of the heart at different points in its pumping cycle, and MRI may prove extremely useful in assessment of infarct and of cardiac output. Even at this relatively early stage of its development MRI of the brain reveals pathology better than any other modality. It is already apparent that MRI of the lumbar spine can serve as a reliable alternative to myelography. Proton MRI appears to be more sensitive than CT in the evaluation of prostatic, bladder, and interuterine carcinomas.

NMR chemical analysis of small volumes of tissue close to the body surface can be accomplished by using the so-called surface coil approach, which defines the region of interest by degrading the static field everywhere else. Much new information about metabolism in intact animals and in humans has already been learned from the proton and phosphorus-31 NMR spectra obtained in this way.

Perhaps the greatest promise of magnetic resonance is to combine imaging and chemical analysis, that is, to provide maps of specific metabolites within the body. Not only would this represent another major advance in quantitative diagnosis, but possibilities for new understanding of metabolic diseases abound. To achieve such chemical imaging it is necessary to separate the spatial and chemical information that are both encoded in the resonance frequency. The separation of water and fat proton images of the body has already been demonstrated. The much lower concentrations of interesting metabolites such as adenosine triphosphate (ATP) and the low signal strength of phosphorus resonances are the challenges that must be overcome to bring chemical imaging to its full power. In the meantime, MRI will continue its rapid pace of clinical applications development with further advances in image interpretation as well as the introduction of such aids as contrast agents.

It is already apparent that clinical research MRI installations require personnel with training in physics. It may well be that within a few years medical physics will be associated with MRI to a greater extent than it is with therapeutic radiology.

PHOTONICS AND MEDICINE

Lasers

The modern age of photonics was heralded by the invention of the laser and the development of various implementations during the early 1960s. Encouraged by the utility found for the radiation sources contributed by physics during the previous several decades, physicians were quick to try laser sources. Because of their acquaintance with light as a factor in function, disease, diagnosis, and therapy, the dermatologists and ophthalmologists were the first to recognize the potential uses of the laser. Because suitable commercial lasers were not generally available then, laser physicists were necessarily involved in much of the pioneering work. Indeed, today many units involved in the development of new laser-based medical procedures include physicists.

Between 1962 and 1964 the continuous-wave argon-ion and carbon dioxide lasers, the lasers of greatest medical utility today, were developed at Bell Laboratories. By 1967 scores of reports and a book describing experimental and clinical procedures that utilized lasers had been published. These writings recognized many of the features of lasers that are used in the field today.

With the possible exception of far-ultraviolet lasers, the essential action of high-power lasers on tissue is localized heating that results in cutting (surgery) and/or cautery. The laser light is absorbed in a small volume of tissue in which the energy is quickly converted to heat. Direct contact of tissue by the laser instrumentation is not required, and the depth and confinement of the laser action can be controlled. Heating of tissues to temperatures between 75 and 100°C even for short periods leads to protein denaturation followed by coagulation and collapse of tissue microstructure. One result is hemostasis, a well-known effect of heat since ancient times. When the temperature exceeds 100°C vaporization of tissue water can occur. As the laser power and dose are increased, greater amounts of steam are produced without much further increase in temperature. If sufficient steam is generated rapidly within tissue, physical separation, or cutting, occurs. Tissues bordering the cut are heated sufficiently by dissipated thermal energy to effect cautery and concomitant hemostasis. Tissues heated beyond complete water vaporization are ablated and carbonized.

The high-power, small focused spot size and wavelength specificity

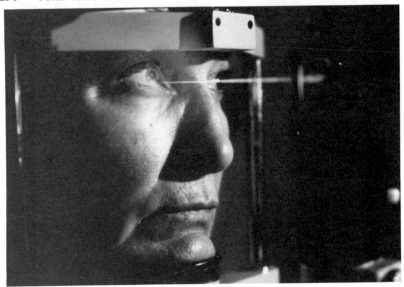

FIGURE 13.3   Argon laser retinal reattachment. (Reprinted with permission from *P&S*, the journal of the College of Physicians and Surgeons of Columbia University, Vol. 4, No. 1, Spring, 1984, p. 30. Photograph by Rene Perez.)

of laser sources coupled with the specific optical properties and heat-transfer properties of various tissues have yielded a variety of medical procedures. Blue light from an argon-ion laser is absorbed strongly by blood and other pigments and is used, for example, for treatment of abnormalities in blood vessels in the skin. Removal of the so-called port wine stain and other birthmarks that are due to hypervascularization is achieved through laser coagulation of the capillary bed in the dermis. With careful study of the physics of the situation, workers have found suitable laser-power levels and pulse widths to achieve capillary coagulation with little damage to surrounding tissue.

Green argon-laser light can pass through the colorless cornea, lens, and vitreous humor but is strongly absorbed by the retinal pigments. Figure 13.3 shows a patient undergoing retina reattachment, one of the oldest medical applications of lasers. Focused high, power bursts of near-infrared light from the yttrium-aluminum-garnet (YAG) laser have been found to be ideal for ophthalmological uses, such as lens cataract removal.

The best laser knife is the focused carbon dioxide laser beam. Because of the strong absorption of the radiation of the carbon dioxide laser by water and the relative inefficiency of heat transfer in tissue, cutting can be much more precisely controlled with the laser than it is with the scalpel. Precision is also improved because the surgeon can view the surgery site continuously during the cutting and through the use of auxiliary optics and aiming beams (helium-neon laser). The carbon dioxide laser is used in gynecology for the removal of pervasive cervical cancer tissue and in fallopian tube surgery, in neurosurgery (opened skull) for removal of deep-seated brain tumors, and in otolaryngology for surgery on vocal cords. Besides increased surgical precision, laser surgery offers reduced bleeding, increased sterility, and much less trauma.

In addition to advances in laser science, physicists have been continuously involved in providing means for delivery of the laser energy to tissue sites. An important contribution was that of the light knife developed in 1967 by physicists D. Herriot and E. Gordon. This device guides the beam from the source through a hollow jointed arm (prisms are located in the joints) to a fountain-pen-sized probe that is held like a scalpel. Flexible hollow waveguides have recently been introduced into carbon dioxide laser-surgery methodology. At present considerable progress has been made in development of a flexible fiber waveguide for 10-μm radiation. Such a guide would permit remote surgical procedures to be performed in internal sites, for example, the colon, the urinary tract, and the bile duct, that can sustain a dry field.

Silica fiber-optic guides are used to guide argon-ion and YAG laser energy. One important application of the guided argon laser beam is in the coagulation of bleeding stomach ulcers. Several procedures that employ fiber-optic-guided laser energy are under development. Perhaps the most exciting of these, laser angioplasty, aims to remove arterial occlusions (clots and arterosclerotic plaque). A group at Stanford University has performed successful laser angioplasty in a clogged femoral artery. The eventual aim is to perform laser angioplasty in coronary arteries. A system for such a procedure requires much development in the ability to view, steer, and aim inside small arteries. Physicists are working with physicians on these problems at several centers.

Fiber Optics in Endoscopes and Sensors

Besides the delivery of energy for therapeutic purposes, there are two other major applications of fiber optics in medicine: endoscopy and sensing. In these uses the information content of the guided light is the focus. In endoscopy, optical images are transmitted from inside the body through a single light pipe or (a more sophisticated technique) through a coherent bundle of ultrathin optical fibers. In fiber-optic sensors light is guided from a source outside the body to a transducer attached to the fiber inside. In response to some chemical or physical property of its immediate environment the transducer alters the incoming light and returns it through the same or a second fiber to photodetectors and analyzers that interpret the signal. The development of both fiber-optic sensors and modern fiber endoscopes depends critically on advances in waveguide optics and integrated optics. Physicists have made large contributions to the recent developments in diode sources and detectors and work on nonlinear optical properties of materials, technologies crucial to the development of fiber-optic sensors.

A large variety of endoscopes is available for visual examination of the interiors of body cavities and hollow viscuses. While their main uses have been for examination of the alimentary canal and urinary tract, they have utility in otolaryngology, gynecology, and pulmonary protocols. Fiber endoscopes for examination of the sinuses of the knee have greatly improved the ability to assess problems there.

Endoscopes range in size from 15 to 0.5 mm and are often combined with devices for biopsy and surgery. Newly available are 0.5-mm-diameter coherent bundles with more than 10,000 4.2-μm-diameter fibers. Such bundles are flexible and are being used in attempts to view occlusions in coronary arteries during catheterization procedures.

Recently, imaging through fiber optics has been combined with fluorescent labeling of abnormal tissue. For example, a derivative of the red-fluorescing dye hematoporphyrin localizes in malignant tumors. Small tumors in the bladder and lung can be detected after intravenous infusion of the dye and a search with endoscopes equipped to bring in blue light and to image any red fluorescence.

Although fiber-optic sensors are too new to have proven value in medical applications, they offer several advantages that ensure their wide use within the next decade as they pass clinical trials. We note the following three advantages: (1) Safety: No electrical connections or electrically conductive materials are involved. Individual sensors are a fraction of a millimeter in diameter and flexible so that multiple sensors can be incorporated in a small catheter. (2) Possibilities for broad application: Many of the in vitro colorimetric and fluorimetric blood tests devised by the clinical chemists can be directly translated into the formulation of transducers for fiber-optic sensors. Optical physicists have described many materials that can report on physical parameters such as temperature and pressure. (3) Simplicity: Because most sensors have relatively simple structures (transducer plus fiber), they should be relatively easy to fabricate and are likely to be inexpensive and disposable.

Recent reviews of fiber-optic sensors list devices for various measurements of the blood including volume flow rate (dye dilution); cardiac output (dye dilution); blood-flow velocity (Doppler effect); temperature, pressure, pH, blood gases (oxygen and carbon dioxide); hemoglobin oxygenation status; and even glucose concentration. There are reports of sensors for calcium-, potassium-, and sodium-ion concentrations as well as optical devices for radiation dosimetry at tumor sites. Several companies are planning to introduce soon multichannel optical devices for continuous monitoring of intravascular blood pressure, blood gases, and blood electrolytes.

Fluorescence Immunoassay

The past decade has witnessed the virtual replacement of the clinical laboratory technician skilled in hands-on wet chemistry by microprocessor-controlled automated clinical analyzers for all the routine blood and urine analyses and many more specialized tests. For example, one portable instrument called the hematofluorometer enables an exceedingly simple and rapid screening test for lead intoxication to be performed on a small drop of blood obtained from the pinprick of a finger. All these modern advances follow the general impact of the microelectronics and photonics revolutions. More recently, quite novel approaches to clinical chemical analysis, especially those based on antibody specificity, have been introduced that represent a direct influence of molecular physics, in particular, of molecular spectroscopy.

Antibodies or antigens of interest are labeled with small molecules that absorb visible light and subsequently emit light at another wavelength region. Unlike the single-event radioactive label, the luminescent label can be repeatedly excited and repeats its reporting, so that the general approach is theoretically more sensitive than RIA (see above). In one new method the polarization of the emitted light (orientation of the light wave) with respect to that of the exciting light is measured. A labeled antigen is a relatively small molecule, and it rotates in random directions fast enough between excitation and emission to depolarize the luminescence (random orientation). If the antigen binds to the antibody, its rotational motion is slowed and the luminescence is polarized. Measurement of the degree of polarization is used to calculate the degree of antibody binding.

In resonance excitation transfer a photoexcited molecule (donor) can transfer its excitation to a different molecule (acceptor) if the second molecule has the appropriate spectroscopic properties and is within a critical distance. A complete theory of resonance excitation transfer was developed in the 1960s by the German physicist T. Förster. This theory allows one to chose appropriate donor-acceptor pairs for a specified transfer range. In recently commercialized immunoassays based on excitation transfer, antibodies are labeled with a fluorescent donor and an antigen with an appropriate acceptor. When antibody and antigen combine the donor fluorescence is quenched.

CLOSING REMARKS

This disquisition on the contributions of physics to medicine was confined to examples in which there was a rather direct application of the physics in the medicine. Even so, many important items had to be left out. Unmentioned, for example, was therapeutic radiology. Neither were subjects such as laser-flow cytometry and implantable auditory neuroprostheses addressed. Noninvasive imaging modalities based on infrared light (for breast tumor detection) and on microwave frequencies near 1000 MHz (for whole-body imaging) are in early stages of development. Ion- and molecule-specific field-effect transistors are being developed for continuous invasive monitoring of a number of metabolites and as components of artificial organs such as an artificial pancreas. Finally, prosthetic devices, including artificial limbs and joints as well as experimental cochlear implants and artificial hearts, have been omitted entirely. The development of such devices owes much to physical materials research.

Of the items that were included, some were cases in which medical application followed fast on discovery of physical phenomenon and some were cases in which medical applications required additional technological development in other areas also derived from physics. In most cases of the latter sort it was the modern microelectronic data processor that provided the enabling technology. Another role being played by the microprocessor in the rapidly changing practice of medicine is in patient management. The electronic patient record combined with expert systems for aiding in diagnosis and disease management will soon bring great benefit in increased diagnostic and monitoring accuracy and in cost savings.

While the modern imaging modalities afforded by advances in physics have contributed significantly to diagnostic accuracy and to the monitoring of the condition and comfort of patients during the diagnostic phase in a cost-effective manner, there is a question concerning the effect of these advanced-technology diagnostic methods on outcome. Diagnostic capabilities in the areas of cancers, cardiovascular disease, and metabolic diseases appear to have outstripped therapeutic capabilities. However, the same sophisticated new diagnostic tools afford the means to follow and evaluate therapeutic modalities. Thus, the rapid advances in noninvasive diagnostic methods of the past decade are showing signs of bringing advances in therapy in the next.

It is highly unlikely that physicians in the next two decades will enjoy the kinds of medical tools portrayed in science fiction. However, fast-developing technological trends responding to economic and institutional forces point to major enhancements of the variety and sophistication of diagnostic as well as therapeutic procedures that are carried out in the physician's office and in private or neighborhood clinics. Automated, accurate, programmable desktop clinical analyzers for routine blood and urine tests are already appearing. Simple and inexpensive instrumentation and reagent kits for batteries of more-complicated analyses associated with clinical specialties are sure to follow. Ultrasound instruments dedicated to cardiovascular or abdominal assessments can be made appropriate for physician's office use. Suitable x-ray and/or NMR scanners designed for use on the head, extremeties, and breast are conceivable. If necessary, digitized images could be transmitted to an expert for real-time consultation. The physician could have immediate access to a lifetime patient record including key diagnostic images as well as to data bases, expert systems, and current epidemiological information. Real-time access to diagnostic and monitoring data for hospitalized patients could also be available. Catheter-based therapeutic procedures, including percutaneous ones, may reduce trauma and relieve the need for the life-support devices associated with current surgery to make many procedures capable of being performed outside the hospital. Much of the technology required to achieve this physician's office of the future is already possible. However, it is evident that much still needs to be done and that fundamental and applied physics will need to contribute at several levels (among them instrument components, molecular understanding, and quantitative physiology). This is certainly true if the practice of medicine is to progress beyond even the concepts mentioned above.

RECOMMENDATIONS

Continued efficient development of diagnostic, therapeutic, and prosthetic instrumentation and devices such as those described above requires maintenance of effective communication and collaboration between research-oriented physicians and physical scientists. There is concern currently that the number of research-oriented physicians, especially those capable of this sort of interdisciplinary interaction, is decreasing below the minimum desirable number. The reasons for this appear to be twofold: the financial rewards of clinical practice are so much less than those of private practice that a research career is considered by many to present an economic penalty, and the difficulties associated with mastering and keeping pace with new instrumentation and device technologies are considered insurmountable. Because of this trend and the trend toward routine use in clinical practice of sophisticated equipment, whose use requires multidisciplinary training

● It is imperative that premedical school and residency programs substantially upgrade training in mathematics and physics.

● A program aimed at reversing this growing gap between traditional education and training in the use of the modern devices that we have described above should be included in the National Institutes of Health training grant structure.

● The number of joint M.D./Ph.D. programs that stress the physical sciences, such as the joint MIT/Harvard program, should be increased.

The record clearly shows that most innovation in medical instrumentation since the turn of the century, even that of the past two decades, has come from the universities and medical schools and not from the medical-device industry. Encouragement of the kinds of academic activities that have brought this progress would therefore increase the rate of innovation in medical devices.

● Encouragement should include strong support of medically oriented applied-physics research programs by the various government granting agencies. The appropriate National Institutes of Health and National Science Foundation organizations should coordinate funding considerations in this arena so that an optimum overlap of physics and medicine ensues.

Index

257

Defects in materials, 71-72, 76, 79-80,
    149-150
  clusters of, 149
Dendritic
  patterns, 85-86, 88
  solidification, 86
2-Deoxyglucose, 243
Diagnostic radiography, 238-239
Diamond pressure cell, 106
Dilution refrigerator, 193-194
Directed-energy weapons, 226-227
Disks
  floppy and rigid, 155-156
  video, 173-174
Disordered materials, 81-85
DNA double helix, 29, 32, 35
Doppler systems, pulsed, 244, 245
DOSECC (Deep Observation and Sam-
    pling of the Earth's Continental
    Crust), 105

E

Earth
  mantle of, *see* Mantle *entries*
  remote sensing of, from space, 106
  as thermodynamic engine, 93-94
Earthquakes, 101, 103
EBL (electron-beam lithography), 140
Ecology, 213-214
Education, recommendations in, 66-67
EELS (electron-energy-loss spectrosco-
    py), 78, 80
Electromagnetic waves (EMWs), 225,
    226
Electron
  accelerators, 17
  materials, organic, 63-66
  microscope, 40, 80
  paramagnetic resonance (EPR), 36, 37
  states, localization of, 82-83
  transport, 32
Electron-atom collisions, 114-115
Electron-beam lithography (EBL), 140
Electron-energy-loss spectroscopy
    (EELS), 78, 80
Electron-hole plasma, 200
Electronics, 14
Elementary-particle physics, 110-111
EMWs (electromagnetic waves), 225, 226
Endoscopes, fiber optics in, 251-253

Energy, 102-104, 197
  environment and, 18-20
  future developments in, 207
  geothermal, 103-104
  nuclear, 103
  physics and, 197-207, 214-215
Engineering, 4
  band-gap, 166
  genetic, 34
Environment
  energy and, 18-20
  physics and, 19, 208-214, 215
Epilayers, growth of, 149
EPR (electron paramagnetic resonance),
    36, 37
Eruptions, volcanic, 101-102
EXAFS (extended x-ray absorption fine-
    structure) spectroscopy, 34, 36,
    61, 78, 80
Executive Summary, 1-2
Extended x-ray absorption fine-structure
    (EXAFS) spectroscopy, 34, 36, 61,
    78, 80

F

FELs (free-electron lasers), 219-220
Fermi National Accelerator Laboratory
    (FNAL), 188
Ferroquinone, 36
Fiber bundles, 173
Fiber-optic
  endoscopes, 251-253
  guides, silica, 251
  sensor technology, 172-173
  sensors, 77
Fictive temperature, 84
Field
  ion microscope, 191
  theory, 126-129
Fission energy, 198
Flame propagation and extinction, 206
Fluid
  dynamics, geophysical, 9
  flows, 118-119
  turbulence, 130
Fluid-mosaic model of cell membrane,
    40-41
Fluids, physics of, 118-119
Fluorescence
  immunoassay, 253-254
  spectroscopy, 37-38

FNAL (Fermi National Accelerator Laboratory), 188
Focused-ion-beam systems, 140-141
Fossil fuels, 102, 197-198
Fractals, 37, 81
Free-electron lasers (FELs), 219-220
Fuels, fossil, 102, 197-198
Funding
  recommendations for, 67-68
  restructuring of, 51
Fusion energy, 198, 202-205
Fusion reactors, radiation effects in, 203-204
Fusion-reactor designs, 201

G

Gallium arsenide (GaAs), 142-145
Gas-phase molecular spectroscopy, 55
Gases, theory of, 130
Gating mechanisms, 45
Gauge theories, 111, 127-128
Gene manipulation, 6, 27
Genetic engineering, 34
Geochemical
  cycles, 97
  reservoirs, 95-97
Geodesy, 92
Geophysical
  data sets, 107
  fluid dynamics, 9
  turbulence, 97-98
Geophysics, 8-10, 91-107
  applications, 101-105
  hazards, study of, 101-102
  research, 91
    future directions of, 105-107
  solid-earth, 92, 107
Geothermal energy, 103-104
Geyser geothermal energy field, 103
Glass-fiber light-guide transmission medium, 167
Glasses, 82
  computer simulation of, 83-84
  metallic, 82
  structural relaxation of, 84
Global digital seismic array, 105-106
Global Positioning System (GPS), 106, 221, 222
Government, industrial development and, 23-24

GPS (Global Positioning System), 106, 221, 222
Grain boundaries, 79, 203
Gravitation theory, 110, 120
Greenhouse effect, 210-212
Groundwater hydrology, 212
Gyroscopes, ring-laser, 218, 219
Gyrotrons, 219, 220

H

Heavy-ion collisions, 117
Helix, DNA double, 29, 32, 35
Hemoglobin, 34
Hemostasis, 249
Heterojunctions, 151
Heterostructures, 144
Higgs fields, 111
High-energy neutron studies, 60
High-pressure studies, 106
High-temperature superconductors, 201
Himalaya mountain belt, 95
Holograms, 174
Hybrid physics codes, 116
Hydrospheric studies, 212-213
Hypersurfaces, potential energy, 58

I

IC (integrated circuits), 136, 137, 139
Ignition processes, 206
Immune system, human, 48
Immunoassay, fluorescence, 253-254
Independent research and development (IR&D), 232, 234, 235
Industrial development, government and, 23-24
Inertial-confinement fusion, 204-205
Information
  age, 134
  network, societal, 13-14
  processing, optical, 173-176
  technologies, optical, *see* Optical information technologies
Information-processing power, 14
Infrared spectroscopy, 38
Instantons, 128
Instrumentation, 16-18, 54, 185-196
  materials science, 76, 80
  medical, 256
  in university research laboratories, 195-196

Proteins, 29, 30
  membrane, 35
Pulsed Doppler systems, 244, 245
Pulsed-laser processing, 200

Q

QCD (quantum chromodynamics), 111,
  127, 128
Quantitative models, 81
Quantum
  chromodynamics (QCD), 111, 127, 128
  electrodynamics, 110
Quantum Hall effect, 143
  fractional, 195
Quark theory, 111

R

Radiation
  damage, 80
  effects in fusion reactors, 203-204
Radioactive dating, 96
Radioecology, 214
Radiography, diagnostic, 238-239
Radioimmunoassay (RIA), 239
Radiology, 237-243
Raman scattering, 38
RBS (Rutherford backscattering spectros-
  copy), 138
Reactive ion etching (RIE), 141
Recommendations, 22-25
  for funding, 67-68
Relaxation
  rates, slow, 90
  structural, of glasses, 84
Renormalization-group theory, 73, 81
Research
  academic, 67
  independent, and development
    (IR&D), 232, 234, 235
  intertwining of basic and applied, 70
  science and, 13
Reservoirs, geochemical, 95-97
Resonance excitation transfer, 254
Restriction enzymes, 35
Restructuring of funding, 51
Retinal reattachment, argon laser, 250
RIA (radioimmunoassay), 239
Ridge push process, 94
RIE (reactive ion etching), 141

Ring-laser gyroscopes, 218, 219
Rutherford backscattering spectroscopy
  (RBS), 138

S

San Andreas Fault, 95, 101
San Fernando earthquake, 101
Satellite technology, 228
Scanning
  electron microscope (SEM), 191-193
  tunneling microscope (STM), 58, 193,
    194
Schottky barriers, 150
Science, research and, 13
Scientific
  interfaces, x
  synergy, 4-11
Security, national, *see* National security
Seismic
  array, global digital, 105-106
  studies of continents, 105
Seismicity, 101
Seismography, 9
SEM (scanning electron microscope),
  191-193
Semiconductor
  lasers, 164-165
  physics, 77
Semiconductors, amorphous, 14, 82, 147
Sensors, fiber optics in, 251-253
Silica fiber-optic guides, 251
Silicon
  solar cells, 199
  technology, 142
Single-crystal neutron diffraction, 34
Smectic liquid crystals, 42
Societal information network, 13-14
Soil physics, 213
Solar
  electrical power, 18
  energy, 198
Solar-energy conversion, 198-201
Solid-earth geophysics, 92, 107
Solid-state creep processes, 94
Solidification front, 87
Solitons, 48, 182-183
Space
  data acquisition from, 104
  missions, 92
  remote sensing of Earth from, 106
Spectral hole burning, 175-176

2464